Climate Change Ethics and the Non-Human World

This book examines from different perspectives the moral significance of non-human members of the biotic community and their omission from climate ethics literature.

The complexity of life in an age of rapid climate change demands the development of moral frameworks that recognize and respect the dignity and agency of both human and non-human organisms. Despite decades of careful work in non-anthropocentric approaches to environmental ethics, recent anthologies on climate ethics have largely omitted non-anthropocentric approaches. This multidisciplinary volume of international scholars tackles this lacuna by presenting novel work on non-anthropocentric approaches to climate ethics. Written in an accessible style, the text incorporates sentiocentric, biocentric, ecocentric, and posthuman perspectives on climate change.

With diverse perspectives from both leading and emerging scholars of environmental ethics, geography, religious studies, conservation ecology, and environmental studies, this book will offer a valuable reading for students and scholars of these fields.

Brian G. Henning is a professor of philosophy and environmental studies at Gonzaga University. He is founding Co-Chair of the climate action group 350 Spokane. His research includes more than 35 articles and nine books, including *Riders in the Storm: Ethics in an Age of Climate Change* and the award-winning book *The Ethics of Creativity*.

Zack Walsh is Research Associate at the Institute for Advanced Sustainability Studies (IASS) in Potsdam, Germany. He co-leads the A Mindset for the Anthropocene (AMA) project, which is a transdisciplinary research project and emerging network of change agents integrating personal and socio-ecological transformations to sustainability.

Routledge Research in the Anthropocene
Series Editors: Jamie Lorimer and Kathryn Yusoff

The Routledge Research in the Anthropocene Series offers the first forum for original and innovative research on the epoch and events of the Anthropocene. Titles within the series are empirically and/or theoretically informed and explore a range of dynamic, captivating and highly relevant topics, drawing across the humanities and social sciences in an avowedly interdisciplinary perspective. This series will encourage new theoretical perspectives and highlight ground-breaking interdisciplinary research that reflects the dynamism and vibrancy of current work in this field. The series is aimed at upper-level undergraduates, researchers and research students as well as academics and policy-makers.

Hope and Grief in the Anthropocene
Re-conceptualising Human – Nature Relations
Lesley Head

Releasing the Commons
Edited by Ash Amin and Philip Howell

Climate Change Ethics and the Non-Human World
Edited by Brian G. Henning and Zack Walsh

Climate Change Ethics and the Non-Human World

Edited by Brian G. Henning
and Zack Walsh

Routledge
Taylor & Francis Group

LONDON AND NEW YORK

First published 2020
by Routledge
2 Park Square, Milton Park, Abingdon, Oxon OX14 4RN

and by Routledge
605 Third Avenue, New York, NY 10017

Routledge is an imprint of the Taylor & Francis Group, an informa business.

First issued in paperback 2021

British Library Cataloguing-in-Publication Data
A catalogue record for this book is available from the British Library.

Library of Congress Cataloging-in-Publication Data
A catalog record for this book has been requested

ISBN 13: 978-0-367-40610-3 (hbk)
ISBN 13: 978-0-429-35698-8 (ebk)
ISBN 13: 978-1-03-223829-6 (pbk)

Typeset in Times New Roman
by Apex CoVantage, LLC

Contents

Contributors

Robin Attfield is Emeritus Professor of philosophy at Cardiff University, Wales, UK, where he has taught philosophy since 1968, and became one of the earliest generations of environmental philosophers with his book *The Ethics of Environmental Concern* (1983 and 1991). His recent books include the textbook *Environmental Ethics: An Overview for the Twenty-First Century* (2003 and 2014); *The Ethics of the Global Environment* (1999 and 2015); and *Wonder, Value, and God* (2017), while his *Environmental Ethics: A Very Short Introduction* appeared in the Very Short Introduction series of Oxford University Press in 2018.

Patrik Baard has a PhD in philosophy from the Royal Institute of Technology, Stockholm, Sweden. He is currently a postdoctoral researcher at the Swedish Biodiversity Centre, Uppsala, Sweden.

Whitney A. Bauman is Associate Professor of religious studies at Florida International University. His books include *Religion and Ecology: Developing a Planetary Ethic* (Columbia University Press 2014) and with Kevin O'Brien and Richard Bohannon, *Grounding Religion: A Fieldguide to the Study of Religion and Ecology, 2nd Revised Edition* (Routledge 2017).

Claudio Campagna is a conservation biologist with the Wildlife Conservation Society. He has been working in species-driven conservation projects for three decades and played a central role in the creation of open ocean marine protected areas. Together with Daniel Guevara, he started the Language of Conservation project under the Center for Public Philosophy at the University of California, Santa Cruz.

Eileen Crist is Associate Professor in the Department of Science, Technology, and Society at Virginia Tech. She is the author of *Images of Animals: Anthropomorphism and Animal Mind*. She is also co-editor of a number of books, including *Gaia in Turmoil: Climate Change, Biodepletion* and *Earth Ethics in an Age of Crisis*; *Life on the Brink: Environmentalists Confront Overpopulation*; *Keeping the Wild: Against the Domestication of Earth*; and *Protecting the Wild: Parks and Wilderness, the Foundation for Conservation*. She is the author of numerous academic papers and popular writings. Her work focuses on biodiversity loss and the destruction of wild places and pathways to halt these trends. Her latest book, *Abundant Earth: Toward an Ecological Civilization*, was published by University of Chicago Press in 2019.

Laura Fernández is a PhD candidate and teacher in training in communication at the Universitat Pompeu Fabra (Barcelona). She has a BA in social and cultural anthropology (Autonomous University of Madrid) and an MA in international studies on media, power, and difference (Pompeu Fabra University, Barcelona). Her research areas are strategic visual communication, critical animal and media studies, and environmental communication ethics. Laura is a member of the Critical Communication Research Group (CritiCC) and a research assistant of the Centre for Animal Ethics in Pompeu Fabra University in Barcelona. She is also the author of a book, *Hacia mundos más animales* (*Towards More Animal Worlds*) published in 2018 by Editorial Ochodoscuatro.

Jeremy Gordon is an assistant professor of communication studies at Gonzaga University. Situated at the intersections of ecocultural theory, environmental rhetoric, critical animal studies, and new materialism, Dr. Gordon's research focuses on questions of how expansive and vibrant multi-species relations are, and might be, constituted. He is particularly invested in how popular culture informs ecological imaginaries and what it means to live in the Anthropocene.

Karen Green has taught philosophy at many Australian universities, on a wide variety of subjects, including environmental philosophy, existentialism, feminism, political philosophy, and philosophy of language. She currently holds a research fellowship at the University of Melbourne. She is the author of *A History of Women's Political Thought in Europe, 1700–1800* (Cambridge University Press, 2014); *A History of Women's Political Thought in Europe, 1400–1700*, with Jacqueline Broad (Cambridge University Press, 2009); *Dummett, Philosophy of Language* (Polity, 2001); and *The Woman of Reason* (Polity, 1995).

Daniel Guevara is Chair of philosophy at University of California, Santa Cruz. He has published mainly in ethics and Kant's philosophy. He has a longstanding interest in Wittgenstein's philosophy and, with conservation biologist Claudio Campagna, formed the Language of Conservation Project, whose purpose is to work out a Wittgensteinian approach to, what Aldo Leopold called, "values as yet uncaptured by language."

Brian G. Henning is Professor of philosophy and Professor and Chair of environmental studies at Gonzaga University. With Rebecca Macmullan and Fawna Slavik, he co-founded the climate action group 350 Spokane. He is the author of more than thirty-five articles and chapters and has authored or edited nine books, including *Riders in the Storm: Ethics in an Age of Climate Change*. His 2005 book, *The Ethics of Creativity*, won the Findlay Book Prize from the Metaphysical Society of America.

Rebekah Humphreys is a lecturer in philosophy at the University of Wales, Trinity Saint David. Her research interests include applied ethics (especially animal ethics and environmental ethics) and moral philosophy in general.

Connie Johnston is a professional lecturer in the Department of Geography at DePaul University in Chicago. Her research interests center on human relationships with the rest of the biophysical world and how those relationships are mediated by Western science. She has several publications related to farm animal welfare science in the US and Europe and the role of scientific language in shaping attitudes toward nonhumans. She is the co-editor of *Humans and Animals: A Geography of Coexistence* (2017).

Eric Katz is Professor of philosophy and Chair of the Department of Humanities at the New Jersey Institute of Technology. A founding officer of the International Society for Environmental Ethics, he has published over 60 articles in the field since 1979. He is the author of *Anne Frank's Tree: Nature's Confrontation with Technology, Domination, and the Holocaust* (2015) and *Nature as Subject: Human Obligation and Natural Community* (1997).

Sam Mickey is an adjunct professor in the Environmental Studies program and the Theology and Religious Studies department at the University of San Francisco. He is the author and editor of several books on ecological philosophy and the environmental humanities, including *Coexistentialism and the Unbearable Intimacy of Ecological Emergency* (Lexington, 2016).

Amanda M. Nichols is a PhD candidate in the Department of Religion at the University of Florida and the managing editor of the *Journal for the Study of Religion, Nature and Culture (JSRNC)*. Her current research and interests include women in the environmental movement and feminist, environmental, and animal ethics.

John Nolt is a distinguished service professor in the Philosophy Department at the University of Tennessee and a research fellow in the Energy and Environment Program of the Howard H. Baker Jr. Center for Public Policy. He has written or edited seven books, three on logic, four on environmental ethics.

Holmes Rolston III, is a university distinguished professor and professor of philosophy emeritus at Colorado State University. His recent book is *A New Environmental Ethics: The Next Millennium for Life on Earth*. Rolston was laureate for the 2003 Templeton Prize in Religion; gave the Gifford Lectures, University of Edinburgh, 1997–1998; and has lectured on seven continents.

Zack Walsh is Research Associate at the Institute for Advanced Sustainability Studies (IASS) in Potsdam, Germany. He co-leads the A Mindset for the Anthropocene (AMA) project, which is a transdisciplinary research project and emerging network of change agents integrating personal and socio-ecological transformations to sustainability.

Foreword

I've witnessed a sea change in the reception of climate change, teaching the subject at a major Virginia university over the last 20 years. (Undoubtedly, many others have had a similar experience.) From a topic eyed suspiciously by students and skeptically by colleagues, the climate crisis has become the looming reality for all.

Yet the pattern of 20 years ago continues to hold on another front. Most people within the academic and broader culture continue to be unlearned about, and oblivious to, the extinction crisis that is unfurling rapidly on multiple fronts: extinction of species and subspecies, extinction of populations and individuals, extinction of ecologies, and mass extinction. Not only is ignorance pervasive, but most people are unable to see that there is an acute justice dimension (both non-human and intergenerational) to this crisis. With the exception of environmental and animal studies programs and the odd philosopher, most academics have neither tools nor wherewithal to think of extinction in a justice framework. Extinction remains a vague problem, and, in my experience, the topic often draws blank stares.

But climate change has become all the rage. Its discourse is a useful lens to discern how the plight of the more-than-human world keeps getting pushed out of view, second-tiered, or mis-framed. Climate change is a clear and present danger to humanity at every level—from individual lives and livelihoods to looming mass displacements of people to imperiling economic life, food production, and trade interconnectivity to endangering industrial civilization as a whole. Unsurprisingly, the sheer fact of climate change monopolizes attention, often to the point of single-issue obsession. Mass extinction, in contrast, having no definitively foreseeable consequences for humanity (though of course it will), is not viewed or framed as an "existential threat." Hence, it is regularly bypassed—most certainly, thus far, by the world's politicians. When extinction is acknowledged along with climate change as a paramount impact that must be addressed, it is mentioned secondarily to climate change and often misrepresented as a consequence of climate change. Widespread ignorance notwithstanding, climate change is not driving the extinction crisis. Agriculture and killing are the main drivers, which really amounts to food production being the main driver since killing terrestrial animals (for bushmeat or to protect livestock) and killing marine animals (for seafood and as bycatch) are tied to food and food markets. (The food system also contributes a hefty proportion of anthropogenic greenhouse gases, roughly 30–35 percent.) So while the extinction crisis is beholden to dominant patterns of land and ocean abuse, this is gingerly skirted and the spotlight directed at climate change.

Undoubtedly climate change is adversely impacting the non-human world and non-humans (and will do so with mounting speed and violence) but mostly in conjunction with the fact that habitat destruction and fragmentation, extinctions of hundreds of thousands of non-human populations, and enormous contractions of historic ranges of species have already mortally weakened the natural world. To drive the point home with a metaphor, if we think of climate change as a "disease vector," it is afflicting non-humans because their "immunity" has been seriously compromised. Earth's biota would be able to handle a certain degree of climate change were it not for the fact that it has been wrecked or wounded by other direct assaults. Even coral reefs, which are exceptionally vulnerable to the current heating event, are doing much better in places of the world where they are undisturbed by other anthropogenic factors. (Beyond the point I am making here that climate change is set to pummel an already damaged world, this fact about coral reefs means we need to strictly protect all of them right now, in order to help them and the biodiversity they harbor cope with coming temperature and acidification spikes.)

The obsession with climate change as looming catastrophe tends to encourage its construal as "the major problem," which is a category mistake. Climate change is a symptom of a no-limitations and increasingly expanding human enterprise, whose reigning ideologies, institutions, and social structures resist even the notion of downscaling humanity's presence and activities in the ecosphere. Framing climate change as The Problem—a view also seeping into the mainstream establishment—is potentially quite expedient for that establishment. This framing leaves the underlying engine of ecological devastation—Earth's colonization—untouched and unscrutinized. Climate change qua "the problem" steers the focus toward technological solutions such as renewables, nuclear power, and climate geoengineering, and their needed (and likely) deployment in some combination or other. The dominant framing of climate change is understandable, since left unaddressed it will lurch the planet into a hothouse state, where massive upheavals and devastations for most complex life will render everything else moot. We should not, however, forsake our capacity for deep analysis in the face of such a menacing possibility. On the contrary, we need to become more incisive and more critical in our thinking about the source of the catastrophe that is already here.

To discern the catastrophe that is already here, picture that climate change is meaningfully addressed within the next two decades by an international community that finally gets its act together to avert mega-disaster(s) for humanity and civilization. (It is too late to avert climate disasters as such, since we are already living through them, and there's more warming in the pipeline no matter what.) Without deeper changes, this will enable business as usual—meaning human dominance within and domination over the natural world—to continue more or less undisturbed except fueled by a "green" energy regime. The extinction crisis will not be resolved by containing climate change, for the simple reason that climate change is not causing it. Human economic, demographic, and techno-infrastructural expansionism is causing it—an expansionism undergirded by the absence of any sense or concept of inequity toward the more-than-human world; an expansionism whose mode of operation toward ecologies, wild non-humans, and farm animals falls mostly in the range between amoral-indifferent and callous-cruel. The dominant human establishment (including, the politically right, left, and center) does not know how to think about inclusive justice and does not know how to imagine a world that is not dominated by humanity. Such an establishment, freed from the threat of Hothouse Earth, would, by the same token, be freed to continue to exercise Earth colonization.

Our situation in this historical moment calls us to look more at the character of our existential presence and less at the symptoms that threaten our existence. After all, if a big asteroid strike occurred, it would wipe out most complex life, including humans. (Thanks to Jupiter containing dangerously sized asteroids in its gravitational field, such strikes occur rarely and are extremely improbable.) The point being that how we live on Earth matters more deeply than how long we exist as a life form in the universe. We can lay no claim to virtue or greatness if we wipe out life's variety, abundance, complexity, and full evolutionary potential—if we let mass extinction happen under our watch and despite its peerless documenta-tion. This is neither my "opinion" nor an ecocentric "perspective." It is reality—the reality of saddling humanity with genocide that will be mirrored in perpetuity outwardly to the universe and inwardly in the human soul. We must not allow this reality to come to pass.

There is a solution to our predicament, one that will preempt mass extinction, soften the blows of climate change, and carry all people through the 21st-century bottleneck more safely and with less suffering. We must downscale our economic, demographic, and infrastructural presence, and we must pull back from and pro-tect vast swathes of nature. To recognize the beauty of scaling down and pulling back, we have to reimagine who we are, where we inhabit, and how we live, while foregrounding goodness toward all beings as the guiding principle in our rela-tions. We need a revolution in insight about who we want to be and how to belong lovingly within the community of life; only secondarily do we need technological renewable transitions.

Eileen Crist

Introduction

Brian G. Henning and Zack Walsh

As the world finds itself mired in social and ecological crises in the opening decades of the 21st century, it is interesting to consider how philosophers responded
to the challenges at the start of the last century. For instance, consider the presidential addresses of the American Philosophical Association (APA) of the 1930s
(1931–1940).[1] While grappling with the diverse philosophical themes represented
in the thirty-odd addresses, it is worth reflecting on the dramatic events unfolding
during the tumultuous decade.[2] At the beginning of the 1930s, the world was sinking into the depths of the Great Depression, with a full quarter of all wage-earners
in America unemployed. And by the end of the decade, virulent nationalist movements in Germany, Italy, and Japan had provoked the Second World War. Juxta-
posing these crises with the addresses of the presidents of the APA is as revelatory
for what was said as what was not.

On the one hand, the addresses delivered at the start of the decade—including
those by prominent philosophers such as Alfred North Whitehead (1931–32) and
C. I. Lewis (1933–34) —make not even passing reference to the global economic
crisis.[3] Though the images of bread lines no doubt featured in their morning
papers may have affected them greatly, the state of global capitalism apparently
did not merit philosophical discussion. The deafening silence regarding the Great
Depression stands in contrast to many of the addresses in the closing years of
the decade, which often focused on growing geopolitical threats.[4] For instance,
in his 1937–38 Western Division presidential address, "History as the Struggle
for Social Values," J. A. Leighton boldly argues for the moral responsibility of
philosophers to become engaged, as philosophers, in the defense of democracy.

> Democracy is emerging in one of the most momentous secular crises in the
> history of culture. On us teachers falls, I think, a heavy share of the burden
> of democracy. If the confusion continues, and some rabble-rouser, greedy
> for power, arises, we may all be either shut up or shot . . . If philosophers
> remain content to pursue the owlish habit of reflection only in the soft even-
> ing twilight of abstract speculation, they are in for a long and dreadful night.
> Civilization is poised on a razor edge over an abyss. To say that we Ameri-
> cans have no responsibility and freedom of choice in this hour of decision is
> a counsel of cowardice and despair . . . I believe that you and I and all of us

are confronted with the challenge to choose whether we shall knuckle down to tyranny controlled by demonic powers or march forward with resolute wills towards the dawn of an earth of worthy and comradely persons – a world set free.

(539–540)

Many of the addresses of the second half of the 1930s similarly reveal that the struggle between democracy and tyranny was as much a war of ideas as of warring armies. And the presidents of the APA were calling on their fellow philosophers to enlist.[5]

This disparity in the philosophical responses to the twin crises of the 1930s naturally invites comparison with the philosophical addresses of more recent APA presidents in the context of our current crises. The greatest existential threat of the 21st century is not economic calamity or the march of tyranny—though these remain all-too-real problems—but global warming.[6] Anthropogenic global climate change is a challenge the magnitude and nature of which the human species has never confronted.[7] At stake is not merely a nation, ideology, or even economic well-being, but the very fate of our species and millions of other species with which we have evolved. What sort of philosophical responses to this crisis do we find in the presidential addresses of the APA?

Given the content of their philosophical addresses, it would seem that the 21st-century presidents of the APA are no more philosophically concerned with global climate change than Whitehead and Lewis were with the Great Depression. Indeed, if the content of these addresses were one's only source, it would be nearly impossible to know that the Earth is in the midst of an accelerating ecological crisis. Amongst the 36 addresses delivered since 2003, not a single APA president makes more than a passing reference to global warming or global climate change.[8] Though philosophers are not blind to its significance, it would seem that too few consider climate change relevant to their philosophical work or their work relevant to climate change. Stephen Gardiner's 2004 observation that "Very few moral philosophers have written on climate change"[9] remains stubbornly true.[10] "This is puzzling, for several reasons." Gardiner continues,

> First, many politicians and policy makers claim that climate change is not only the most serious environmental problem currently facing the world, but also one of the most important international problems per se. Second, many of those working in other disciplines describe climate change as fundamentally an ethical issue. Third, the problem is theoretically challenging, both in itself and in virtue of the wider issues it raises. Indeed, some have even gone so far as to suggest that successfully addressing climate change will require a fundamental paradigm shift in ethics.
>
> (Gardiner, "Ethics and Global Climate Change" 555–56).

To be sure, as Gardiner observes in the note attached to his claim, several philosophers—including most notably John Broome, Dale Jamieson, and Henry

Shue—have been writing on climate change since the early 1990s. However, philosophers in general and moral philosophers in particular have been relatively slow to focus their efforts on climate change. But there are signs of progress.

Thanks in good part to the work of Gardiner, Jamieson, and others, since 2004, there has been considerable philosophical attention paid to climate change, with important anthologies on ethics and climate change being published by Oxford in 2010 and Cambridge the year following, as well as others.[11] What is noteworthy from the perspective of the current project is that these climate ethics projects are not an outgrowth of the rich and diverse field of environmental ethics but of mainstream Anglo-American moral philosophy. A chief implication of this fact is that, to date, the work done on the ethics of climate change has been almost exclusively anthropocentric in its philosophical orientation. This anthropocentric philosophical orientation toward climate ethics is all the more surprising given that, as Jamieson notes in 1992 and repeats in 2004, successfully addressing climate change will require a fundamental paradigm shift in ethics (Jamieson 1992: 292; Gardiner, "Ethics and Global Climate Change" 555–56). Jamieson rightly notes that

> our dominant value system is inadequate and inappropriate for guiding our thinking about global environmental problems, such as those entailed by climate changes caused by human activity. This value system, as it impinges on the environment, can be thought of as a relatively recent construction, coincident with the rise of capitalism and modern science. . . . It evolved in low-population-density and low-technology societies, with seemingly unlimited access to land and other resources. This value system is reflected in attitudes toward population, consumption, technology, and social justice, as well as toward the environment.
>
> (Jamieson, Ethics, Public Policy, and Global Warming 148)

Though as diverse as any other subdiscipline of philosophy, the critique of "dominant value systems" has been one of the longstanding motivations of many within environmental ethics.[12] Despite their philosophical differences, many philosophers take a primary (if not the primary) goal of environmental ethics to construct a more adequate conceptual framework and axiological foundation for thinking about the nature of moral obligation. Given this, it is surprising that major anthologies on climate ethics have generally not included non-anthropocentric approaches to climate ethics.[13]

The primary goal of the present volume is to bring attention to and begin to address this deficiency in contemporary climate ethics literature by bringing together a variety of non-anthropocentric approaches to climate ethics. Self-consciously pluralistic, this volume brings together both leading and emerging scholars from not only environmental ethics but also fields as diverse as geography, religious studies, conservation ecology, and communication studies. The discussion begins with philosopher **John Nolt's** essay, "Climate Change and the Loss of Nonhuman Welfare," which describes how the biotic and hedonic welfare of non-humans can be objectively measured, granting an overall better picture of the health of individuals and species outside anthropocentric lenses. The well-being and suffering of different species is, of course, sometimes incomparable or incommensurable, especially across different types of welfare; nevertheless, aggregate accounts of non-human welfare can be assessed and compared to shed greater light on the alarming rates of welfare lost and the ways we can redouble our conservation efforts.

In "Anthropocentrism and the Anthropocene: Restoration and Geoengineering as Negative Paradigms of Epistemological Domination," philosopher **Eric Katz** characterizes ecological restoration and geoengineering as prime examples of human epistemological domination—the idea that the non-human environment should be shaped by how we understand its value and purpose. Katz shows how scientific and technical approaches to environmental control and management delegitimize strategies for environmental protection and mitigation. He defines artificial environments as environments shaped by human intentionality and decries the decimation of natural environments, as well as the hubris of naming our current geological epoch the Anthropocene, after ourselves. Katz concludes by advocating not for preserving pristine nature but for respecting ecological processes and their capacity to heal themselves.

From the perspective of communication studies, in the third chapter, "Climate Ethics Bridging Animal Ethics to Overcome Climate Inaction: An Approach from Strategic Visual Communication," **Laura Fernández** critiques anthropocentrism and explains how its prevalence in climate ethics and visual communications leads to the suffering of non-human animals. She uses examples from the agribusiness and aquaculture industries to illustrate how denying the suffering of farmed animals is linked to climate policy's neglect of non-human lives. She explores how visual communications can strategically represent non-human animals to change people's attitudes and behavior. Though not every strategy she presents is non-anthropocentric, Fernández's chapter connects the fields of animal and climate ethics in practical ways to better acknowledge the value of non-human lives.

The focus on animal ethics is continued in Chapter 4, "Suffering, Sentientism, and Sustainability: An Analysis of a Non-Anthropocentric Moral Framework for Climate Ethics," where philosopher **Rebekah Humphreys** documents how different versions of sentientism influence practical responses to climate change and species loss, in so far as they incorporate the intrinsic value and interests of individual, sentient non-humans. Though admittedly more inclusive than anthropocentrism, Humphreys finds sentientism less inclusive and holistic than bio- and eco-centrism which attribute moral significance to species groups and the systems that support them. Since sentientism does not acknowledge the intrinsic value of non-sentient creatures, Humphreys claims, it fails to protect them, in turn failing to better serve even sentient beings, whose well-being depends on the flourishing of non-sentient creatures.

Making the transition to biocentric approaches fully explicit, in "Biocentrism, Climate Change, and the Spatial and Temporal Scope of Ethics," philosopher **Robin Attfield** analyzes various accounts of anthropocentric ethics, finding them logically indefensible, untenable, or in conflict with each other. By contrast, non-anthropocentric ethics, Attfield says, reflect human ethical responsibility toward non-human species near, far, and undiscovered. Only biocentric and ecocentric ethics, he claims, are able to reflect our moral responsibility toward issues of climate change and species loss, both of which are distributed across vast spatial and temporal scales. Ultimately, Attfield defends biocentrism as the more practicable ethics, concluding with a call to advance education programs that cultivate biocentrism in public life.

Conservation ecologist **Claudio Campagna** and philosopher **Daniel Guevara** provide another biocentric approach in their collaboration entitled "Evaluating Climate Change with the Language of the Forms of Life." Arguing that the relative lack of public concern for the future of biodiversity is partly due to environmental discourses describing non-human welfare in instrumental terms, as natural capital or ecosystem services, they develop an alternative biocentric framework using the concept of "natural goodness" advanced by philosophers Michael Thompson and Philippa Foot. This offers a basis for the kind of radical reorientation in values required for institutional change to help alleviate species extinctions.

Following this prescriptive proposal, from the perspective of religious studies, **Whitney A. Bauman** advocates three changes to the university that will help create a planetary rather than an anthropocentric understanding of our world in "Thinking Through the Anthropocene: Educating for a Planetary Community." First, science and religion should be understood together, rather than as separate entities. Second, the university should decolonize epistemology by paying attention to other than Western systems of knowledge, including the wisdom and perspectives from the more-than-human world. And third, the university should reimagine human beings, first and foremost, as planetary creatures among other creatures, rather than as citizens of a specific culture, race, and religion.

In the eighth chapter, philosopher **Patrik Baard's** essay "Conflicting Advice: Resolving Conflicting Moral Recommendations in Climate and Environmental Ethics" distinguishes claims made by environmental ethics and climate ethics in the context of policy responses to climate change and biodiversity. He analyzes how environmental and climate ethics view moral conflicts arising from various Intergovernmental Panel on Climate Change (IPCC) scenarios, based on recommendations to implement negative emissions technologies, limit population growth, enhance technological efficiency, and decrease consumption. Baard concludes by calling for the development of a climate ethics that is informed by environmental ethics, drawing attention to the benefits of cultivating virtues like humility and charity.

Philosopher **Karen Green's** essay, "An Eco-centric Proposal for Setting a Price on Greenhouse Gas Emissions," considers the failures of previous attempts to regulate global carbon emissions. She documents the problems inherent in regulating emissions on a per-capita basis. Although nations with disproportionately large populations share greater responsibility for reducing emissions, distributing responsibility accordingly fails to address higher consumerism in developed countries, developing countries' desire to further develop, and the global need for promoting alternative low-carbon intensive development. As such, Green proposes regulating emissions and distributing responsibility for reductions based on a nations' or regions' landmass relative to its capacity for sequestering carbon. Though it's an imperfect measure, she argues, it is a more just alternative that can be feasibly implemented with enough political will.

Religious studies professor **Amanda M. Nichols's** chapter "Being Human: An Ecocentric Approach to Climate Ethics" describes ecocentrism as a challenge to the predominance of anthropocentrism in Western science, philosophy, and religion. She characterizes ecocentrism in terms of several general criteria, arguing that it is non-anthropocentric, upholds the intrinsic value of non-human nature, critiques the economistic view of nature as a resource, advocates population reduction, and views nature holistically, prioritizing wholes over parts. Nichols's essay then evaluates the recent writings of three posthumanists (Donna Haraway, Edwardo Kohn, and Timothy Morton) to examine areas of dialogue between ecocentrism and posthumanism.

Whereas Nichols views Haraway and Kohn much more favorably than Morton, environmental studies professor **Sam Mickey** provides an extended defense of Morton's position and its relevance for climate change ethics in his essay "Atmospheres of Object-Oriented Ontology." Mickey illustrates how Morton's thought, and object-oriented ontology (OOO) in general, leads to a particularly unusual non-anthropocentric ethics. Unlike many climate ethics, which focus on the pragmatics of our response, Mickey focuses on the affective condition of being conditioned and why any response to climate change inevitably fails to satisfy. He contends that there is nevertheless joy amidst such despair. By plumbing the depths of our conditionality, we playfully attune to others, human and non-human, on their own terms.

Returning again to a communication studies lens, **Jeremy Gordon's** "Monsters, Metamorphoses, and the Horror of Ethics in the 'Pelagioscene'" develops posthuman perspectives on our relationship to watery bodies using illustrations from two films, "Splash!" and "The Shape of Water." His essay focuses Mickey's theoretical discussion of strange human and non-human affectations and affiliations via feminist-inspired reflections on porosity, vulnerability, and humility. By becoming intimate with the monsters in and around us, Gordon offers speculative visions of our condition in the Anthropocene, elucidating our problematic relationship to toxic environments, both social and ecological.

This analysis of climate ethics and posthumanism is continued in "Gut Check: Imaging a Posthuman 'Climate'" in which geographer **Connie Johnston** deconstructs the unitary human subject at the center of the Anthropocene by exploring how the "human" is co-constituted by both macrological and micrological climates. Johnston theorizes what it means to be human in light of recent scholarship on posthumanism and gut microbiota, illustrating that those quintessentially human characteristics, like rationality, are functions of ecological relations, affecting and affected by anthropogenic climate change. Discarding the foundational logics of humanism, she reinscribes the human in relation to the non-human, following the logics of posthumanism to illustrate ways of advancing non-anthropocentric ethics.

The volume concludes with a soaring essay by one of the intellectual founders of environmental ethics, **Holmes Rolston III**, in "Wonderland Earth in the Anthropocene Epoch." Here, Rolston describes the exceptional cosmological and evolutionary conditions that gave rise to the flourishing of Earth, life, and humans. He explains how evolution accelerated as the human brain allowed us to develop civilization via language, thought, and large-scale cooperation. We now live in a geological epoch, called the Anthropocene, in which humans are the preeminent geophysical force shaping the Earth system. Rolston documents how the pervasiveness of human influence has led many scholars to proclaim the end of nature and the need to further control non-human life through geoengineering schemes. Rolston, however, responds by situating the human in symbiotic relation to the non-human, emphasizing greater humility and a more holistic approach to appropriately scaled development.

The chapters in this volume are diverse in their disciplinary, methodological, and theoretical perspectives. Nevertheless, this project is far from exhaustive in representing the breadth and depth of non-anthropocentric approaches to climate change. For instance, missing here are deep ecology and ecofeminist approaches. Also missing are indigenous and non-Western perspectives, philosophical or otherwise. More work is needed to gather and focus these important voices as our species seeks to understand and respond to the existential threat of global climate disruption and mass species extinction.

To take more than a modicum of literary license, let us conclude this brief introduction by imagining what J. A. Leighton might say if he delivered his presidential address today not on the threat of tyrants but on the obligation of philosophers in the face of climate catastrophe:

> Global climate change is emerging as one of the most momentous crises in the history of our species. On us teachers falls, I think, a heavy share of the burden of responding to the crisis. If the thirst for unnecessary consumption continues, we may all be either inundated or extinct. . . . If philosophers remain content to pursue the owlish habit of reflection only in the soft evening twilight of abstract speculation, they are in for a long and dreadful night. Much of life on Earth is poised on a razor edge over an abyss. To say that we philosophers have no responsibility in this hour of decision is a counsel of cowardice and despair. I believe that you and I and all of us are confronted with the challenge to choose whether we shall knuckle down to the tyranny of anthropocentrism or march forward with resolute wills towards the dawn of an Earth community in which all forms of life are respected – a biotic community set free.

Notes

1 All of the presidential addresses of the APA were published in a special series with one volume dedicated to each decade of the 20th century, for a total of ten volumes. (See *American Philosophical Association Presidential Addresses*. In the 11th volume, the editor, Richard Hull, invited scholars to write a new essay inspired by and responding to a particular decade of presidential addresses. I was asked to write on the 1930s. (See

Henning, "Philosophy in the Age of Fascism: Reflections on the Presidential Addresses of the American Philosophical Association, 1931–1940."

Note, the first paragraphs here are an updated version of the opening to my "From the Anthropocene to the Ecozoic: Philosophy and Global Climate Change," published in *Midwest Studies in Philosophy*, 284–95.

2 Philosophy is rather odd in having not only three divisions (now called the Pacific, Central, and Eastern) but also three presidents, one for each division. This means that within one decade, there are usually thirty presidential addresses.

3 For a list of past presidents and their addresses, see www.apaonline.org/?page=presidents.

4 As I note in "Philosophy in the Age of Fascism," "Although from our twenty-first century vantage point, the eventual victory of the Allies might seem to have been inevitable, in the closing years of the 1930s, Japan had invaded and committed terrible atrocities in China, and Hitler and Mussolini were on the march across Europe. Eventual victory was not so certain then, and the existential threat to democracy was palpable in many of the addresses delivered during this period" (88).

5 See especially the addresses by James H. Tufts (Pacific Division President 1934–1935), "The Institution as an Agency of Stability and Readjustment in Ethics"; E. T. Mitchell (Western Division President 1935–1936), "Social Ideals and the Law"; J. A. Leighton (Western Division President 1937–1938), "History as the Struggle for Social Values"; Glen R. Morrow (Western Division President 1939–1940), "Plato and the Rule of Law"; and Edward O. Sisson (Pacific Division President 1939–1940), "Human Nature and the Present Crisis." Note that the "Western Division" has since been renamed the "Central Division."

6 Decades-long stagnant wages culminating in the speculation-fueled Great Recession reveals deep and likely systemic problems with the global capitalist economic system. In many ways, just as the Great Depression may be seen as a primary cause of World War II, it is likely that the excesses of capitalism are a primary driver of anthropogenic climate change. See Naomi Klein, *This Changes Everything: Capitalism vs. the Climate*. Daly and Cobb's *For the Common Good* is also an important resource. Philip Cafaro is among the philosophers who have forcefully called for the end of the pursuit of economic growth. See, for instance, Cafaro, "Taming Growth and Articulating a Sustainable Future," 1–23. For a positive alternative to the current growth model, see Herman Daly and John B. Cobb Jr. *For the Common Good*. This paper is in agreement with these positions, though it does not enter into the debate.

7 Gardiner's work on climate change as a "perfect moral storm" is helpful here. Gardiner, "A Perfect Moral Storm: Climate Change, Intergenerational Ethics, and the Problem of Corruption," 87–98. See also Dale Jamieson, "There are three important dimensions along which global environmental problems such as those involved with climate change vary from the paradigm: Apparently innocent acts can have devastating consequences, causes and harms may be diffuse, and causes and harms may be remote in space and time" (Jamieson, "Ethics, Public Policy, and Global Warming," 149).

8 A keyword search of "global warming" and "climate" of the 42 presidential addresses delivered between 2003 – 2017 reveals only three fleeting references: Nicholas Smith, "Modesty: A Contextual Account" (2007–2008 Pacific Division presidential address) (26) www.apaonline.org/global_engine/download.asp?fileid=91467BC7-9407-4D84-B53B-56188A221F01); Linda Martín Alcoff (2012–2013 Eastern APA presidential address) "Philosophy's Civil Wars" (35) www.apaonline.org/global_engine/download.asp?fileid=DB5D29E5-9DC5-4D3D-A4BC-3550069C3147); and Elizabeth Anderson (2014–2015 Central APA presidential address) "Moral Bias and Corrective Practices: A Pragmatist Perspective" (23) www.apaonline.org/global_engine/download.asp?fileid=93FF3C8F-DDF0-4CDE-BDB2-91B5603BE414).

9 Gardiner, "Ethics and Global Climate Change," 555.

10 The inactivity of the APA is in contrast to, for instance, the American Academy of Religion (AAR). Much of the 2014 annual meeting of the AAR was dedicated to the topic of climate change, including a session with former President Jimmy Carter on

"The Role of Religion in Mediating Conflicts and Imagining Futures: The Cases of Climate Change and Equality for Women" and plenary panels with, among others, the then chair of the Intergovernmental Panel on Climate Change (IPCC), Rajendra K. Pachauri, and 350.org founder Bill McKibben. More to the point, Linda Zoloth's presidential address was dedicated to the topic: "Interrupting Your Life: An Ethics for the Coming Storm."

11 See *Climate Ethics: Essential Readings; The Ethics of Global Climate Change*. Slightly more popular readers have also been published, such as *Moral Ground: Ethical Action for a Planet in Peril*, and *The Global Warming Reader*.

12 The most significant counterexample to this would be the so-called environmental pragmatists. See in particular Andrew Light and Eric Katz, *Environmental Pragmatism*.

13 Gardiner et al.'s *Climate Ethics* does not include any non-anthropocentric approaches to climate ethics, unless you include utilitarian approaches, which are sentiocentric in their orientation. Arnold's *The Ethics of Global Climate Change* fares somewhat better, including one non-anthropocentric essay out of fourteen, see Clare Palmer on "Does nature matter? The place of the nonhuman in ethics of climate change," 272–91.

References

Arnold, Denis G., ed. *Climate Ethics: Essential Readings; The Ethics of Global Climate Change*. Cambridge: Cambridge University Press, 2011.

Cafaro, Philip. "Taming Growth and Articulating a Sustainable Future." *Ethics & the Environment* 16, no. 1 (2011): 1–23.

Daly, Herman E., and John B. Cobb Jr. *For the Common Good*. Boston: Beacon Press, 1994.

Gardiner, Stephen M. "Ethics and Global Climate Change." *Ethics* 114, no. 3 (2004): 555–600.

Gardiner, Stephen M. "A Perfect Moral Storm: Climate Change, Intergenerational Ethics, and the Problem of Corruption." in *Climate Ethics: Essential Readings*, 87–98. Oxford: Oxford University Press, 2010.

Gardiner, Stephen M., Simon Caney, Dale Jamieson, and Henry Shue, eds. *Climate Ethics: Essential Readings*. Oxford: Oxford University Press, 2010.

Henning, Brian G. "Philosophy in the Age of Fascism: Reflections on the Presidential Addresses of the American Philosophical Association, 1931–1940." In *Historical Essays in Twentieth Century American Philosophy, vol. 11 of the American Philosophical Association Presidential Addresses*, general edited by Richard Hull, 11 vols. Charlottesville, VA: Philosophy Documentation Center, 2015.

Henning, Brian G. "From the Anthropocene to the Ecozoic: Philosophy and Global Climate Change." *Midwest Studies in Philosophy* 40 (2016): 284–95.

Jamieson, Dale. "Ethics, Public Policy, and Global Warming." *Science, Technology, Human Values* 17 (1992): 139–53.

Katz, Eric, and Andrew Light. *Environmental Pragmatism*. London: Routledge, 1999.

Klein, Naomi. *This Changes Everything: Capitalism vs. The Climate*. New York: Simon & Schuster, 2014.

McKibben, Bill. *The Global Warming Reader*. New York: Penguin Books, 2011.

Moore, Kathleen Dean, and Michal P. Nelson, eds. *Moral Ground: Ethical Action for a Planet in Peril*. San Antonio: Trinity University Press, 2010.

Palmer, Clare. "Does Nature Matter? The Place of the Non-Human in the Ethics of Climate Change." In *The Ethics of Global Climate Change*, edited by Denis Arnold, 272–91. Cambridge University Press, 2011.

Richard Hull, general ed. *American Philosophical Association Presidential Addresses*, 11 vols. Charlottesville, VA: Philosophy Documentation Center, 2015.

1 Climate change and the loss of non-human welfare

John Nolt

What would it mean to conceive the harms of climate change and related environmental degradation to non-human life not only as biodiversity losses or threats to species, but, more fundamentally, as losses of welfare—losses, in other words, of non-anthropocentric (and non-anthropogenic) goodness? To begin to see, take a long look back.

Good times and bad

There have been times when Earth was richly inhabited and times when it was barren, times when it was molten hot, and times when it was bitterly cold, times when life was ascendant for tens of millions of years, and times of mass extinction. To everything, it is said, there is a season.[1] But some seasons have been better than others.

I don't mean better for *us*. Humans weren't around for most of this. Nor am I thinking of what was good for us in that it set the stage for our existence or in that it was the sort of world we would have preferred—for its grandeur, say, or hospitableness—if we were there. Leave us and our preferences and predilections out of it. That is what it means to think non-anthropocentrically. I mean times when there was much that was good or much that was bad independently of us.

The period following the Chicxulub asteroid strike about 66 million years ago, for example, was a bad time. That impact and its consequences—a monstrous tsunami, worldwide wildfires, and climatic disruptions—eliminated three-quarters of all species. All non-avian dinosaurs died. Suffering, starvation, injury, and death were terrible beyond comprehension.

Times since then, however, have been comparatively beneficent. Biodiversity has blossomed. Innumerable beings in fantastic variety have arisen and flourished—and, even though (as evolution demands) most have perished prematurely, many have reproduced. The sentient ones have felt pain or joy or something analogous to them.

Think, too, of times when nothing was either good or bad because there was nothing for which things could be good or bad—the billion or half-billion years, for example, between the Earth's formation and the advent of microscopic life. During that entire period, nothing—on this planet, at least—was good or bad for anything. It is still that way in some places—on the planet Mercury, for example. There are mountains, craters, rocks, and dust particles on Mercury, but nothing is good or bad for them. So far as we know, nothing good or bad has *ever* happened on Mercury—though, of course, much has happened. But on Earth there have been good times and bad and times that were neither.

Biologists now worry that by refashioning the world to suit ourselves in the short term, we may be bringing on another bad time of long duration. They don't put it that way, of course, at least not usually. They speak instead of biodiversity loss, threats to species, diminishment of ecosystem services, habitat destruction— and, lately, of a threatened sixth mass extinction.[2] (The Chicxulub event was the fifth.) There are those—as Holmes Rolston III notes in his contribution to this volume—who celebrate the escalating human power that is inflicting these losses. But understood non-anthropocentrically, without the blinders of short-term and exclusively human values, they are overwhelmingly harmful. Unchecked, they will add up to a very bad time.

Harm is welfare reduction. Benefit, correlatively, is welfare maintenance or increase. Welfare, harm, and benefit can be understood either individually or collectively. Harm, for instance, is individual when a living creature is sickened or wounded or killed. It is collective when the total welfare of a population declines—either through degradation of its members' welfare or loss in their numbers, or both. A "good time" is a period when the collective or aggregate welfare of the living beings on Earth is relatively high and steady or increasing. A bad time is a period when it is relatively low or plummeting.

Three types of welfare

Etymologically, welfare is simply faring well. But the concept is complicated. There are at least three main types of welfare: biotic, hedonic, and a third type that is much rarer.

Biotic welfare is, roughly, physical health. Every living organism—indeed, every living cell—has some degree of it. It can be characterized as autopoietic functioning (Nolt 2009). The basic autopoietic functions are encoded in the organism's genetic script. Among them, for all organisms, are obtaining and metabolizing nutrients, eliminating wastes, regulating hydration, and so on. But the biotic welfare of an organism is not merely the totality of its autopoietic functions; it is also their mutual integration into the organism's self-sustaining goal-directed behavior. The quantity of an organism's biotic welfare is the quantity of its biological life.

That quantity is always positive or zero, never negative. Zero biotic welfare is death.[3] When an organism is ill or injured, its biotic welfare is still positive but lower than normal. When it is healthy, its biotic welfare is relatively high. Complex organisms are capable of more biotic welfare—more life—than simple ones, because they have more living subsystems and higher levels of functional integration.

Non-sentient organisms have *only* biotic welfare. They fare well just to the extent that they function well as integrated living systems. Thus, they can be harmed only by diminishment of their physical health—reduction, we may say, of their *quantity* of life. They do not consciously experience harm.

But biotic welfare is not the only kind. The nervous systems of many complex, mobile animals have evolved so that some essential functions (e.g., eating, drinking, rest, nurturance, sex) produce pleasure, and certain kinds of damage or potentially damaging conditions (e.g., injury, illness, hunger, thirst, the appearance of enemies) produce anguish or pain. Pleasant experiences evoke repetition; painful ones, avoidance. Thus, there arises associatively learned behavior that is conducive to survival and reproduction.[4] The ability to experience pain and pleasure is *sentience.*

Sentience evolved, no doubt, because it promotes biotic welfare, hence survival and reproduction. But positive and negative feelings have in the process acquired some degree of independence from those original functions. Positive feelings are intrinsically rewarding; negative ones, intrinsically disagreeable. In many sentient animals, these valences have come to constitute a distinct kind of welfare: a welfare of feeling. Call it *hedonic* welfare. Hedonic welfare may be either positive or negative.

Welfare in sentient animals is therefore complex, having both a positive biotic component that may be low or high and a hedonic component that ranges from negative to positive. Hedonic welfare overlays, but does not replace, biotic welfare. Sentient animals, like all living beings, have a degree of biotic welfare—typically a relatively high degree, because of their complexity and high levels of functional integration. But they also have a degree (positive, negative, or mixed) of hedonic welfare. They fare well to the extent that they have high levels of both—to the extent, in other words, that they are physically fit and feeling fine.

The two types of welfare may clash. Sometimes, for example, the misery of a sentient individual is so intense, prolonged, and unrelievable that the individual's overall welfare is no longer positive (though it may not be negative either; see the following section on welfare comparison). That may justify euthanasia.

In addition to biotic and hedonic welfare, there is a third kind: a welfare of meaning that is apparently available only to social animals. This is the kind of well-being enacted, for example, in love or friendship, or in the pursuit and achievement of a career, knowledge, or creative expression. Because it is a social or cultural phenomenon, the welfare of meaning transcends mere feelings. It is prominent for, though not limited to, humans. (Think, for example, of friendships between humans and animals or among animals.) But it is absent from most living things. Since this chapter's topic is non-human welfare, I focus here mainly on biotic and hedonic welfare.[5]

One final—and crucial—preliminary point: biotic and hedonic welfare are *objective* conditions. When a non-human animal suffers from a wound or feels pleasure in the presence of its offspring or mate, that is a fact independent of us, not something we project anthropomorphically into the situation. Such things occurred for many millions of years before we arrived on the scene.

The feeling-lives of animals are, of course, very different from ours, and there is much about them that we can't understand. Still, they can to some extent be investigated empirically, often these days neurophysiologically. Biotic welfare is even more obviously objective. Whether a tree is healthy or diseased, for example, is a fact that we can determine by empirical examination. It is healthy to the extent that it does well and resiliently what trees of its type do (e.g., grow, photosynthesize, draw water and nutrients from the soil). Often, we can tell just by looking.

Welfare and ethics

Cards on the table: I am an ethical biocentrist. I hold that the welfare of living beings of any kind, whether sentient or not, ought to have some moral significance—though for very simple individual organisms that significance is often negligible in practice. That biocentrism motivates this chapter, but having argued for it elsewhere, I won't do so here.[6] Nor will I assume it. What I do assume is that all living beings have some degree of objective welfare and that having positive welfare is good for the living being that has it. This section explains those assumptions.

Philosophers have usually conceptualized welfare in one of three ways: hedonically, as described in the previous section; epithumetically, as satisfaction of desires; and, finally and most recently, as a set (or "list") of objective properties.

Welfare conceived epithumetically has, like hedonic welfare, both positive and negative valences: desire-satisfaction is positive and desire-frustration negative. And like hedonic welfare, the epithumetic conception of welfare is generally thought to be limited to sentient or conscious beings—though at least one attempt (not very successful in my opinion) has been made to generalize the idea to non-sentient life (Agar 2001; Nolt 2015a, 177–78).

Straightforward epithumetic conceptions miss the mark, however, because desire- or preference-satisfaction is poorly correlated with anything that could reasonably be called welfare. This is clearly true for contemporary humans, both because of the poor fit of our instinctive preferences with twenty-first century life and because our cultivated preferences are much manipulated by propaganda and advertising. Getting what we want is often objectively bad for us (consider junk food or addictive substances), or it may leave us bored and vaguely dissatisfied (many consumer products have this effect). Getting what we do not want (the end of a toxic relationship, for example) may be good for us. Sometimes we want too much; our welfare can in that case be improved by reducing rather than satisfying desires. Satisfaction of preferences (or what Agar calls *biopreferences*) is likewise poorly correlated with welfare for non-human animals in non-natural habitats. Red wolves that were reintroduced into the Great Smoky Mountains in the 1990s had preferences for the taste of antifreeze that they found puddled in parking lots. Some of them died in misery as ethylene glycol crystallized in their kidneys.[7]

There are, of course, attempts to modify epithumetic conceptions of welfare, usually by some sort of idealization, but these add interminable complications. It is best not to go down that rabbit hole.[8]

That leaves us with hedonic and objective-list conceptions. *Exclusively* hedonic conceptions have absurd consequences. Noszick's experience machine is the best-known illustration (Nozick 1974, 42–45). This is a thought experiment in which we can maximize pleasure (hedonic welfare) only by remaining always wired to a machine that produces unsurpassably pleasurable illusions. Simple ethical hedonism implies that we *should* stay wired. That, however, is ethically absurd.

Still, those who would refuse the experience machine need not deny that positive feelings contribute to welfare and negative feelings detract from it. Noszick's thought experiment merely shows that feelings are not all there is to welfare.

That brings us to objective list conceptions, which hold that welfare is constituted not by subjective states alone, but by certain objective conditions. In humans, these conditions may include health, security, freedom from poverty and oppression, longevity, education, meaningful work, rich personal relationships, or the like. (There is, of course, disagreement over the list's contents.) Some objective list theories also include enjoyment and the absence of pointless suffering, thus subsuming hedonic welfare into objective welfare.

Objective list theories are broad enough to encompass all three of the welfare types discussed in the previous section: hedonic, biotic, and the welfare of meaning. Biotic welfare (physical health) should certainly be on the list. It is an objective form of welfare, if anything is. Of course, the goodness of health can sometimes be outweighed by, for example, mental torment. But unless some such strongly countervailing negative condition is present, it is always better for an individual to be healthy than not. Hedonic welfare should also be on the list, because pleasure so obviously enhances the quality of a life and suffering equally obviously detracts from it. I therefore assume an objective list theory of welfare whose list, at minimum, incorporates both biotic and hedonic welfare. Objective list theories also typically—and, I think, rightly—include some aspects of the welfare of meaning. But, as noted earlier, the focus of this chapter is on biotic and hedonic welfare.

Most current objective list theories are limited to humans, or to humans and sentient animals. But if physical health (biotic welfare) is on the list, as it should be, then the theory is in principle generalizable to all living beings. That generalization requires the list to be disjunctive (not all listed conditions need be satisfied for some welfare to be present), but some disjunctiveness is necessary even in the human case, since not all humans are capable of all the kinds of welfare that are typically listed.

Broadening our conception of welfare to all living beings does not entail commitment to *ethical* biocentrism. There is no inconsistency in admitting that bacteria or insects have biotic welfare while denying their moral considerability (Nolt 2006). Yet even those who don't regard them as morally considerable may find the study of their objective welfare useful and illuminating. Take, for example, conservation biologists.

Welfare in conservation biology

On the surface, the concerns of conservation biology seem very different from those of value theory. Conservation biologists typically study aggregates or types, not individuals *per se*. They are interested, for example, in how well a *genus* or *species* or *subspecies* of living things is doing. But that is, at least in part, a question about welfare.

The welfare of a taxon—for concreteness sake, let's say a *species*—is, of course, not the same thing as the welfare of its current members, even taken in aggregate. When a pack of wolves kill a caribou, for example, the aggregate welfare of the caribou species is diminished by the loss of this individual. But in the long run, predation benefits the species by culling the old, the weak, and the sick, thus increasing genetic fitness.

This difference between the welfare of the species and the collective welfare of its current members has suggested to some that the species is a kind of "super-organism" with welfare of its own that is distinct from the welfare of its members (Rolston 1988, ch. 4). But this difference is better understood as the result of a shift in modal and temporal perspective. In the caribou case, it is the difference between the *current* aggregate welfare of the species population and the prospect of sustaining reasonably high aggregate welfare for the species population over many generations.[9]

I take Robin Attfield to be advocating a similar view of species welfare in his essay "Biocentrism, Climate Change, and the Spatial and Temporal Scope of Ethics" in this volume. And, incidentally, I agree with Attfield, too, in denying that ecosystems have welfare over and above the welfare of their inhabitants. Attfield puts it this way:

> If . . . ecosystems are understood as their component populations at given times, then adding ecosystems [in tallying welfare] would simply involve double-counting, since all these individuals are taken into account by biocentrism already, and also by the kind of ecocentrism which includes biocentrism.

I would add only that the welfare of an ecosystem, like that of a species, is best understood not as a momentary snapshot but as a trajectory.

Species welfare thus understood is *in part* closely related to the notion of species *status* in conservation biology. The International Union for Conservation of Nature (IUCN) classifies the status of wild species and other taxa into five categories: Critically endangered, Endangered, Vulnerable, Near-threatened, and Least concern. Classification into these categories is determined by the following criteria:

A   Declining population (past, present, and/or projected)
B   Geographic range size and fragmentation, decline, or fluctuations
C   Small population size and fragmentation, decline, or fluctuations
D   Very small population or very restricted distribution
E   Quantitative analysis of extinction risk (e.g., Population Viability Analysis)

Parts of items C and D take current population into account. The other items are all predictors of future population. The IUCN categories are therefore all concerned with population size, including the limiting case of extinction (population zero). For conservation biologists, then, how well a species is doing—its conservation status—comes down operationally to the size of its current population and the likelihood of robust population size in the future.[10]

That, however, is not the *full* concept of species welfare. The aggregate welfare of a species population at a time is a function not only of the population's size and anticipated size but also of the welfares of its individuals. (We may think of it as the projected population times the anticipated average welfare of species members.) Individuals of functionally complex species have, as was noted earlier, greater biotic welfare than individuals of simpler ones. And in sentient species, there is also hedonic welfare to consider.

Such considerations are absent from the concept of species status. They do, however, arise in conservation biology in a less formal, and more controversial, way. It is a cynical commonplace that conservation NGOs focus their money and publicity on "charismatic megafauna." Biologists and ecologists often criticize this emphasis for its neglect of unglamorous species, some of which are much more important ecologically. Certainly, there is some pandering to human prejudices. But there is more to it than that. Because complex animals generally have high levels of welfare, prioritizing them can also reflect a reasonable interest in the levels of the various types of welfare that the animal can have. Our interest in charismatic megafauna is not, therefore, entirely unjustifiable.

Much good was eliminated from the world recently, for example, when the baiji (Yangtze river dolphin, *Lipotes vexillifer*) was driven to extinction by pollution, fishing, and river traffic[11]—good not just for humans (though there was that) but for this unique, sensitive, and intelligent animal. The welfare of the species *Lipotes vexillifer* was not constituted just by its population size, nor by the trajectory of its population size, though these were relevant factors, but also by the welfare levels of its members, which included both biotic and hedonic components—and perhaps (who knows?) some aspects of the welfare of meaning as well.

Porpoises are complex, intelligent creatures. But at the other end of the scale of life are one-celled organisms, whose welfare, as was noted earlier, is merely biotic and utterly minuscule. Given their vast populations, however, their welfare may be substantial in aggregate. And some of these organisms—cyanobacteria, for example—play such important ecological roles that they are essential to the survival, and hence welfare, of much else. Their contribution to the total welfare of life on Earth no doubt exceeds the totality of their own welfare.

Welfare assessments, even for these very humble organisms, might usefully be incorporated into conservation biology.[12] Since the relevant sorts of welfare are objective and open to empirical investigation, this is in principle possible. Moreover, since welfare is a kind of goodness, if only for the individuals that have it, both biology and value theory might be advanced by efforts to track it—regardless of whether that welfare is taken to be morally considerable. Given measures of the biotic and (where relevant) hedonic welfare of individuals, it would be possible to determine an average individual welfare for a species. The aggregate welfare of the species at a given time could then be obtained as the product of that average with the species population at that time. Trends in aggregate welfare could be monitored to provide richly informative estimates of species welfare. That would enhance our understanding of what it means, not just for us, but for all living things, to be headed toward a bad time of long duration. And who knows what might be discovered along the way?

Welfare comparisons: incomparability, incommensurability, and finitude

But is any of this feasible in practice? How, for example, could biotic welfare be measured? The field of public health uses a measure that could serve as a model for conservation biologists. This is the QALY, or quality-adjusted life-year, a measure of human health over time. In the simplest version of the idea, a year lived by a person in perfect health counts as one QALY. Death, the minimal state of health, yields zero QALYs. A year of life in less than perfect health has some value between 0 and 1. There are standard ways of measuring this value.[13] Similar measures could be developed for the biotic welfare (at least) of many non-human species. Presumably, because simpler organisms generally have less biotic welfare than humans, a QALY for one of them would at best be some fraction of 1.

The welfare quantities of living beings probably are not, however, in all cases comparable. There are two reasons for this. First, even considering only non-humans, there are two fundamentally different kinds of welfare: biotic and hedonic. There is no apparent way to combine both measures on the same scale. Second, within each type there is so much diversity that values of the same type still may not be comparable.

Two quantities are *comparable* if one is either greater than or less than or equal to the other. They are *incomparable* if this is not the case. Incomparability has nothing to do with ignorance. It is a matter of the structure of the greater-than relation among the quantities. Where there is incomparability, values are not linearly ordered, but they may still be partially ordered. In a typical partial order, some *are* greater than others. But some are neither greater than nor less than nor equal to some others; they are simply different.[14]

Consider first just biotic welfare. Given the great diversity of life forms, probably there are no uniform and universal criteria for it. It is therefore quite likely that, on any reasonable measurement scheme, some biotic welfare values are incomparable with others. The biotic welfare of a healthy dandelion plant, for example, may be neither greater than nor less than nor equal to that of the healthy earthworm nestled near its taproot. The welfares of these two organisms may just be too different for intelligible comparison. Yet incomparability in some cases does not preclude comparability in others. Between a healthy paramecium and a healthy potato beetle, for example, the beetle, with its greater complexity and higher levels of functional integration, has the higher biotic welfare. Certainly, the biotic welfare of a human is enormously greater than the biotic welfare of a bacterium.

Biotic welfare values, then, are probably only partially ordered. Among biotic welfare values (none of which are negative), zero is less than all others. But values greater than zero need not form a line of increasing value above zero; rather, they may form a network. Hedonic welfare values are likewise almost sure to be merely partially ordered, as I have argued regarding comparisons of suffering across species (Nolt 2013).

Comparison across welfare *types* (e.g., comparison of some quantity of biotic welfare with some quantity of hedonic welfare) is even more problematic. Such quantities are, I assume, not only incomparable but *incommensurable*. Two values of the same valence (positive or negative) are incommensurable when no amount of either equals or exceeds any amount of the other—that is, any amount of either is incomparable with any amount of the other. Thus, for example, it seems reasonable to suppose that no amount of pleasure is either greater than or equal to or less than any amount of health, for there seems to be no common unit in which both could be measured.

Yet even incommensurability does not preclude all comparisons, provided that we regard objective welfare as a single concept with multiple dimensions.[15] Any animal with certain levels of biotic and hedonic welfare, for example, has more welfare than any organism with lower levels of both—and also than any *non-sentient* organism with a lower level of biotic welfare. Likewise, any population with high aggregate biotic and hedonic welfare has greater welfare in aggregate than any population with lower levels of both. Therefore, although multidimensional welfare values are in general merely partially ordered, they can still encode a good bit of potentially useful information. We might, for example, represent the aggregate welfare of a habitat, an ecosystem, and maybe even the entire biosphere as such a multidimensional value. Such a representation might have both ethical and scientific uses.[16]

One might wonder, however, whether the biotic welfares of the simplest organisms are of any practical interest at all. There is reason to think that they are, at least in aggregate, for the welfare of even a single bacterium, though minute, is not *infinitesimal*, relative even to that of a human. In fact, there are no *infinite* welfare differences among living things. This is because any two organisms—on Earth, at least—have a common ancestor, and the number of reproductive steps in each of their lineages from that ancestor, though it may be astronomical, is finite. Furthermore, offspring always resemble their parent (or parents) closely enough so that at none of these steps could the offspring's welfare have differed infinitely from that of the parent(s). (This holds even if the welfare values are multidimensional and their order is merely partial.) Therefore, no welfare difference along any lineage is infinite. But since any two organisms have a common ancestor, it follows that the welfare of neither is infinite relative to that of the other.[17]

Conclusion: climate change and the welfare of life

There have been good times and bad. We know the history. Yet here we are—an allegedly intelligent and knowledgeable species—swarming to the brink of a very bad time. We really ought to stop, and look, and think—and back up.

Anthropogenic climate change can take life on Earth over that brink for three reasons: (1) it is deadly, (2) we have been unable to control the behavior that is causing it, and (3) its effects are of long duration. Consider each of these reasons in turn.

(1) Climate change now claims hundreds of thousands of human lives annually (DARA 2012). That number is growing. Even at current rates, the cumulative total will be in the tens of millions by the end of this century (Broome 2012, 33). Those are deaths. The number of casualties (which includes not only deaths but injuries and illnesses) is much higher (Nolt 2015b, 2018). Climate mortality or casualty estimates are unavailable for most of the living things with which we share this planet, so we do not have even approximate estimates of their climate-change-induced mortality—which cannot easily be distinguished from losses due to other anthropogenic factors, including habitat destruction, pollution, and overexploitation. We do, however, know this: overall population losses are already broad and steep (Ceballos, Ehrlich, and Dirzo 2017), and extinctions are likely to accelerate with increasing temperature (Urban 2015). The trend is clear.

(2) The threat of anthropogenic climate change has been well understood since the 1980s. As I was composing this essay, the *New York Times Magazine* published a stunning interactive article by Nathaniel Rich about the opportunities that we missed back then. Its title was "Losing Earth: The Decade We Almost Stopped Climate Change."[18] Limiting planetary temperature requires large and rapid reductions in global carbon emissions; yet despite the landmark Paris Climate Accord of 2016, emissions during 2017 were not only *not* decreasing but *increasing* at an alarming rate. (The growth that year was 1.6 percent).[19] Decades of effort to reduce the combustion of fossil fuels have thus borne little fruit. Dale Jamieson, who was among the first philosophers to write about the dangers of climate change, subtitled one of his recent books *Why the Struggle against Climate Change Has Failed—And What It Means for Our Future* (Jamieson 2014). It's hard to disagree. We have, so far at least, been unable to control ourselves.

(3) I wrote earlier of tens of millions of human deaths by 2100. But temperatures will not suddenly return to normal in 2100. The best recent estimates of the duration of the warming that our emissions are causing are on the order of tens of thousands of years (Archer 2009; Zeebe 2013). Ocean acidification, also caused by our carbon emissions, is irreversible on roughly the same time scales (IAP 2009). Biodiversity's recovery from the resulting extinctions may take *millions* of years (Barnosky et al. 2011, 51; Kirchner and Weil 2000). These are effects of long duration.

We can conceive all of this abstractly with statistics. But to grasp its significance, we also have to face it concretely, as losses of individual lives and individual welfare—as injury, illness, displacement, and death to living beings. In human terms, this hit home to me in the fall of 2016, when the Tennessee Valley, where I live, was for weeks filled with smoke from wildfires resulting from an unprecedented four-month drought. Fourteen people died in the raging fires in Gatlinburg and Pigeon Forge, and well over a hundred were injured. Many more lost their homes. Those were the human casualties—the losses of human welfare. Add to that the suffering and deaths of sentient animals—deer, foxes, chipmunks, tortoises—in the rich Southern Appalachian forests that were reduced to ashes. Add the loss of the trees, the insects, the understory plants, the fungi, and the microorganisms in the soils that baked for hours—and you get some sense of what non-anthropocentric welfare loss means. Non-anthropocentric welfare—non-anthropocentric *goodness*, really—is a chief source of the wonder of what in this volume Holmes Rolston III rightly calls "wonderland Earth."

Jamieson is right. The struggle against climate change has failed. Climate tragedies are multiplying. Much of this world's goodness is being lost. But what failed was just the opening engagement in what must be a much longer struggle. We must redouble our efforts to keep the remaining fossil fuels in the ground. And the work of saving, healing, and restoring has just begun.

Notes

1 I am alluding, of course, to the song "To Everything There Is a Season" by the great American singer-songwriter and activist Pete Seeger—and hence to Ecclesiastes 3:1–8, from which he borrowed most of the lyrics. The song was popularized as "Turn, Turn, Turn" by the Byrds.
2 See, for example, Barnosky et al. (2011). Kolbert (2014) is a well-written popular account by a Pulitzer Prize-winning journalist.
3 Nolt (2015a, 175–78). For a discussion and defense of the idea that death is zero welfare in the human case, see Bradley (2009, 98–111).
4 Perhaps this occurred several times. The evolution of sentience in squids and octopuses, which are mollusks, for example, apparently happened separately and very differently from its evolution in vertebrates.
5 There are other ways to categorize these concepts. Aaron Smuts (2018), for example, distinguishes meaning from welfare, which he regards as constituted by mental states, thereby neglecting its most fundamental form: health. His concept of welfare is therefore too narrow for the purposes of this chapter.
6 See especially Nolt (2015a, chs. 5–6) and sec. 7.3.1; Nolt (2010, 2016).
7 Guffey (2005, 131–32).
8 I go down it a little way in Nolt (2015a). See the discussion of idealized preference satisfaction theories on p. 55 and the general argument against preference-satisfaction theories on pp. 76–78.
9 For more on how to do this, see Nolt (2015a, 185–89, 2017a).
10 See www.iucnredlist.org/, accessed August 14, 2018.
11 See http://us.whales.org/case-study/baiji-first-dolphin-to-be-declared-extinct-in-modern-times, accessed August 10, 2018.
12 For further thoughts on how this might be done in practice, see Nolt (2017a).
13 See https://en.wikipedia.org/wiki/Quality-adjusted_life_year, accessed August 14, 2018.
14 I develop a formal axiomatic theory of partially ordered value structures in a monograph entitled *Incomparable Values* (in preparation). For an elementary discussion of partially ordered biotic values, see Nolt (2015a, 237–40). Nolt (2013) provides a discussion of partially ordered hedonic values.

15 Such a concept, obtained by identifying the zero-levels of its component dimensions, would be invariable in the sense articulated by Eden Lin (2018, 345). For a detailed example of how this can be done, see Nolt (2017b).

16 For further elaboration of this idea, see Nolt (2015a, 173–83, 237–40, 2017a).

17 This is a crude summary of a technical argument. That argument is laid out in more detail in Nolt (2017b). A fully rigorous and much more general version is contained in my manuscript *Incomparable Values* (see note 14).

18 www.nytimes.com/interactive/2018/08/01/magazine/climate-change-losing-earth.html? rref=collection%2Fsectioncollection%2Fclimate&action=click&contentCollection=cli mate&region=stream&module=stream_unit&version=latest&contentPlacement=20&p gtype=sectionfront, published August 1, 2018, accessed 8/15/2018.

19 *BP Statistical Review of World Energy*, 67th ed., 2, www.bp.com/content/dam/bp/en/ corporate/pdf/energy-economics/statistical-review/bp-stats-review-2018-full-report. pdf, accessed August 15, 2018.

References

Agar, Nicholas. *Life's Intrinsic Value: Science, Ethics and Nature.* New York: Columbia University Press, 2001.

Archer, D., et al. "Atmospheric Lifetime of Fossil Fuel Carbon Dioxide." *Annual Review of Earth and Planetary Sciences* 37 (2009).

Barnosky, Anthony D., et al. "Has the Earth's Sixth Mass Extinction Already Arrived?" *Nature* 471 (2011): 51–57.

Bradley, Ben. *Well-Being and Death.* Oxford and New York: Oxford University Press, 2009.

Broome, John. *Climate Matters: Ethics in a Warming World.* New York: W. W. Norton, 2012.

Ceballos, Gerardo, Paul R. Ehrlich, and Rodolfo Dirzo. "Biological Annihilation Via the Ongoing Sixth Mass Extinction Signalled by Vertebrate Population Losses and Declines." *Proceedings of the National Academy of Sciences* 114, no. 30 (2017): E6089–E6096. doi:10.1073/pnas.1704949114.

Development Assistance Research Associates (DARA). *Climate Vulnerability Monitor.* 2nd ed. 2012. Accessed August 15, 2018. https://daraint.org/climate-vulnerability-monitor/climate-vulnerability-monitor-2012/.

Guffey, Stan. "Biota." In *A Land Imperilled: The Declining Health of the Southern Appalachian Bioregion,* edited by John Nolt. Knoxville: University of Tennessee Press, 2005.

IAP (Interacademy Panel on International Issues). "IAP Statement on Ocean Acidification." 2009. Accessed August 15, 2018. www.interacademies.net/File.aspx?id=9075.

Jamieson, Dale. *Reason in a Dark Time: Why the Struggle Against Climate Change Has Failed—And What It Means for Our Future.* Oxford and New York: Oxford University Press, 2014.

Kirchner, James and Anne Weil. "Delayed Biological Recovery from Extinctions throughout the Fossil Record." *Nature* 404, no. 9 (2000): 177–80.

Kolbert, Elizabeth. *The Sixth Extinction: An Unnatural History.* New York: Henry Holt, 2014.

Lin, Eden. "Welfare Invariabilism." *Ethics* 128 (2018): 320–45.

Nolt, John. "Are There Infinite Welfare Differences Among Living Things?" *Environmental Values* 26 (2017): 73–89.

Nolt, John. "Casualties as a Moral Measure of Climate Change." *Climatic Change* 130, no. 3 (2015): 347–58. doi:10.1007/s10584-014-1131-2.

Nolt, John. "Comparing Suffering Across Species." *Between the Species* 16, no. 1 (2013): 86–104.

Nolt, John. "Cumulative Harm as a Function of Carbon Emissions." In *Climate Change and Its Impacts: Risks and Inequalities*, edited by Colleen Murphy, Paolo Gardoni, and Robert McKim. Cham, Switzerland: Springer, 2018.

Nolt, John. "Die Nachhaltigkeit des Wohls des Lebens" ("Sustaining Life's Welfare"). German translation by Christoph Haar. In *Nachhaltigkeit und Transition: Conzepte*, edited by Rosa Sierra and Anahita Grisoni, 119–40. Frankfurt and New York: Campus Verlag, 2017.

Nolt, John. *Environmental Ethics for the Long Term: An Introduction*. New York and Abingdon: Routledge, 2015.

Nolt, John. "Hope, Self-Transcendence and Biocentrism." In *Ecology, Ethics, and Hope: A Bird in the Gale*, edited by Andrew Brei. London and New York: Rowman and Littlefield, 2016.

Nolt, John. "Hope, Self-Transcendence and Environmental Ethics." *Inquiry* 53, no. 2 (2010): 162–82.

Nolt, John. "The Move from *Good* to *Ought* in Environmental Ethics." *Environmental Ethics* 28, no. 4 (2006): 355–74.

Nolt, John. "The Move from *Is* to *Good* in Environmental Ethics." *Environmental Ethics* 31, no. 2 (2009): 135–54.

Nozick, Robert. *Anarchy, State, and Utopia*. New York: Basic Books, 1974.

Rolston, Holmes, III. *Environmental Ethics: Duties to and Values in the Natural World*. Philadelphia: Temple University Press, 1988.

Smuts, Aaron. *Welfare, Meaning, and Worth*. New York and Abingdon: Routledge, 2018.

Urban, Mark C. "Accelerating Extinction Risk from Climate Change." *Science* 348, no. 6234 (2015): 571–73.

Zeebe, Richard E. "Time-Dependent Climate Sensitivity and the Legacy of Anthropogenic Greenhouse Gas Emissions." *Proceedings of the National Academy of Sciences* 110 (2013).

2 Anthropocentrism and the Anthropocene

Restoration and geoengineering as negative paradigms of epistemological domination

Eric Katz

I

The recent introduction of the concept of the Anthropocene to describe the present and coming age of human life on Earth legitimizes the idea that the human domination of the natural world is the normal state of affairs. This is a form of anthropocentric epistemological domination, in that human thought alone—before the creation of any human policy alternatives—prescribes the subjugation of the non-human natural world as the fundamental idea in the relationship of humanity to nature. The non-human world is treated merely as an instrument for the promotion of human ends, the very essence of anthropocentrism. This anthropocentric framework clearly leaves little or no room for the direct moral consideration of the non-human world.

The acceptance of the concept of the Anthropocene has a direct effect on how we approach the ethical issues raised by the new reality of climate change. For if the world—both human and non-human—is deemed to be an instrument (or collection or system of instruments) for the promotion of human ends, then the existence of climate change becomes a problem of proper and efficient management: how do we best mitigate the worst effects of climate change, or how do we best adapt to the new conditions of a world of climate change, so that the promotion of human interests continues? The consideration of the effects of climate change on the non-human world is only significant for its effect on human life and institutions. The acceptance of the concept of the Anthropocene as the proper description of our reality on Earth thus means that we accept the validity of anthropocentrism as the proper framework for the evaluation of ethics and the development of policy regarding the crisis of climate change. The non-human world has no direct value, and we can (and should) modify it wherever we can to respond to climate change and preserve human goods.

In this chapter, I explore the idea of the human domination of the natural world that is the essence of an anthropocentric framework for understanding the human relationship to the environment. I use the projects of (1) ecological restoration and (2) geoengineering as paradigmatic examples of the human domination of the natural world, as both a physical phenomenon and an epistemological idea. The kind of thinking that is embodied in the policies of both restoration and geoengineering is directly relevant to our understanding of the normative character of the possible human response to climate change. Both restoration and geoengineering are projects derived from the same kind of thinking that labels the current age the Anthropocene, an essential anthropocentrism that considers human interests as the primary (and sometimes, only) end of human action and policy. By examining the anthropocentric foundations of both restoration and geoengineering, we will see plainly the reasons why climate change policy fails to consider the value of the non-human world.

II

For almost thirty years, I have been a steadfast philosophical critic of the process of ecological restoration (see Katz 1992, 1993, 1996, 1997, 2000, 2012, 2015a).[1] In my view, briefly, restoration ecology is a scientific and technological project that requires the human domination of natural processes. Its stated goal is the restoration, repair, and re-creation of damaged and destroyed natural landscapes. Physically, restoration involves the use of human science and technology to mold, manipulate, and manage the natural processes that exist in these areas. Epistemologically, restoration attempts to impose a human ideal on the newly re-created landscapes. Thus, there is both a physical and epistemological domination of the non-human natural world: humanity controls the physical processes as well as the meaning and value of the environment.

There are many variations of the processes of ecological restoration. At one extreme, it is the cleanup and mitigation of local or regional systems damaged by human interference, such as clearing the waste from a nearby stream or (on a larger scale) cleaning up the damage from the 2010 BP oil disaster in the Gulf of Mexico. At the other extreme, it can be an attempt to re-make a natural ecosystem after an intentional disturbance (or even destruction) of the system, such as the re-creation and re-planting of a mountainside after strip-mining for coal. There are, of course, many possible kinds of restoration projects between these two extremes, a broad spectrum of activities that can all be labeled as ecological restoration. One of the earliest examples I cited was the work of restorationist Steve Packard and his attempt to re-create the oak-savannah plains of the American Midwest (Katz 1991; reprinted in Katz 1997). Re-wilding projects in Europe that seek to re-claim abandoned agricultural land (Tanasescu 2017; and see Drenthen 2018) could also be considered as falling in the mid-range of the spectrum of possible cases.

My criticisms of the process of ecological restoration have rested primarily on my claim that the imposition of human design onto a natural system produces a human-based artifact rather than a natural entity or system. The idea that an ecological restoration project actually "restores" a natural system is a misuse of the word *restore*. We cannot restore a natural system; at best, we can create a perfect substitute, but this substitute is an artifact, not a naturally occurring system. Artifacts are different from naturally occurring entities because artifacts are the products of human intentionality and design; naturally occurring entities lack the presence of human intentionality and design. I consider this difference to be an essential ontological difference: artifacts and natural entities are different kinds of beings. Thus, once we impose our intentional design onto a natural system, we no longer have a truly natural system: we have, perhaps, a garden, a farm, or a zoological park. We may also create a landscape that appears to be a natural wilderness because for various reasons we appreciate the aesthetic qualities and experiences of wilderness—but such a wilderness (designed by humans) is an artifact that merely resembles the wilderness produced by natural processes.

As an example of the difference between artifacts that incorporate human intentionality and naturally occurring entities that do not, consider a hypothetical walk that I take through a forest. Just as I am hoping for a brief rest, I see a fallen tree. I sit on the fallen tree as if it were a chair. The fallen tree functions as a chair, for me, in this situation, but it is clearly not a chair. I could sit on this tree for an infinite amount of time, and it will never, by itself, become a chair. Only the addition of human intentionality and action can create a chair from the naturally occurring tree. Artifacts are the result of human intentionality modifying the processes of nature; natural entities and systems are those not guided or modified by human intentions. Thus, the re-creation of restored natural areas and systems through the process of ecological restoration is the human production of artifacts.

Helena Siipi has refined the idea of artifacts in relation to natural entities in a way that supports my general claim (see Siipi 2003, 2008; Katz 2012, 2015a). Emphasizing that these distinctions must be understood along a gradient or spectrum, so that an entity can be more or less an artifact or more or less a naturally occurring entity, Siipi notes that the necessary condition of an artifact is that its "properties . . . have been intentionally modified by a human being or by a group of human beings" (2003, 415–18). But the amount of modification is significant: adding one sunflower to a meadow does not make the entire meadow into an artifact. What is most important is that the intentional actions of the human being bring the entity (artifact) into existence by causing it to have certain properties. This analysis allows us to recognize why a chair is an artifact but removes an intentionally conceived human infant from the realm of artifactually designed beings. (At least at the present time. Siipi's analysis suggests that in some future time, the possible selection of human traits for infants will render them as a type of artifact.) Moreover, this analysis can serve to refute some spurious counterexamples raised against my position, as for example, the existence of a polluted stream. Although intentional human activity caused the stream to be polluted, the intentional activity did not cause the stream to come into existence. On Siipi's account, then, the stream is not an artifact.

Restoration projects, however, are definitely artifacts, since they are entities and systems that come into being because of human intention and design. The restored entities would not exist with the properties that they have without human intention, design, and modification. The artifactual quality of the restoration project—in reality, its essence—reveals the anthropocentric character of the entire process. Human preferences guide the restoration process in that the re-created landscape is a landscape desired by humans. Ecological restoration ecology is an anthropocentric project: it is based on human interests. Indeed, it is a continuation of the paradigm of human scientific and technological mastery over natural processes. The underlying philosophical assumption is that humans can control natural processes to better effect than nature can. Welcome to the optimistic vision of the Anthropocene: the age of human mastery of the planet Earth.

My criticism of the ecological restoration is not, I argue, a mere quibbling over words. There are implications for both environmental policy and philosophy. If humanity believes that its science and technology can restore any natural entity or system, then the environmental policy goals of protection and preservation become meaningless. Control, modification, and manipulation become the means and ends of environmental policy. Government and industry policymakers will be able to argue that anything can be done to a natural entity or system to benefit human interests, because the damaged or despoiled entity or system can be restored later. The idea of preservation of the natural environment will lose all substantive content. In the Anthropocene age, ecological restoration demonstrates that we will live in a world of the unlimited modification of natural entities and processes.

The philosophical conclusion is also clear: human action regarding the natural environment will know no limits. Humans will modify the environment in whatever ways serve human interests. The natural world will have no value except for its usefulness to human projects of control and domination. Since no non-human interests or values are considered, restoration ecology exhibits a form of anthropocentric thought without any limitations: anthropocentric epistemological domination.

III

Recent ideas for geoengineering the Earth as a means of alleviating the worst effects of climate change is an intensified continuation of both the physical domination of the natural world and this anthropocentric thought process that I have termed epistemological domination.[2] It is not at all clear that geoengineering on a planetary scale—as in proposed projects of solar radiation management—will be successful in "solving" the problem of climate change, but if it were, it would mean both the end of any nature that is non-human and the end of modes of thought that seriously consider non-anthropocentric being and value. Almost thirty years ago, Bill McKibben famously declared "the end of nature" because the unintentional addition of massive amounts of carbon dioxide into the atmosphere since the beginning of the Industrial Revolution had fundamentally changed the natural world (McKibben 1989). But plans for geoengineering a solution to climate change are *intentional* forms of planetary change. The resulting planet would be a human artifact, a highly complex machine designed and managed by human-centered interests and goals. As with the local and regional projects of ecological restoration, the value of the planet and the entities that comprised its systems would be measured solely on the basis of satisfying human interests. Thus, the geoengineered planet, just as a restored landscape, would be a product of the anthropocentric domination of nature.

Maialen Galarraga and Bronislaw Szerszynski provide a cogent analysis of the ontological character of artifacts that reinforces my ideas about geoengineering as a form of human domination of the natural world. They distinguish between two types of end-results of a process of making artifacts. There are stable artifacts that exhibit few dynamic properties, such as pieces of furniture; these "endure in a way that depends on minimizing the exchange of material and energy across their boundaries." But then there are metastable artifacts that contain dynamic properties; these artifacts "maintain their existence dynamically through the controlled exchange of material and energy with this environment." Examples of metastable artifacts are "fires, fields, and gardens"—and, of course, climates (Galarraga and Szerszynski 2012, 223).

Galarraga and Szerszynski also distinguish three kinds of activity in the making of the artifacts: production, eduction, and creation. Production is the simplest making, the imposition of a form upon formless matter, as in placing a mold on a lump of clay. Galarraga and Szerszynski note that Hannah Arendt ascribed the relation of "domination" to this creative process, as the material of nature is forced into following a plan or intention of the human fabricator. The second form of making is eduction, wherein we focus on the process by which the final artifact is drawn out of the characteristics of the material. In understanding the making of a brick through the process of eduction, we focus on the quality of the clay to be molded or shaped; moreover, from the perspective of eduction, making is never complete or finished, for the artisan may work through a constant state of revision. The third form of making is creation, an "ontological genesis," that results in an entirely new entity different from any matter and form that preceded it. This act of creation is literally the building of a new world, whether we create a new kind of tool, human institution, or social process (225–227).

These distinctions can be applied to geoengineering and SRM with significant results. What type of making is the deployment of SRM? The idea of production, in which human agency imposes a new complete and final order on the climate process is clearly the model that "dominates the contemporary discourse" (228). Yet this model fails to understand the fact that climate is a metastable artifact that is constantly changing, so a more appropriate model would be eduction, in which the form of the climate would emerge out of the human interactions with the climate, "through recursive learning and adjusting" (230). In the eductive model of the making of the climate, human intentionality will be forced to adjust to the processes, the internal characteristics, of nature. In "geoengineering by SRM [human action] becomes tied to the continuous task of modulating climatic processes and thus subject to their logic" (230). It is a process similar to what I have termed (in a different context) the "imperialism of nature"—natural processes act as if they were imposing their interests upon the human world (Katz 1995; reprinted in Katz 1997).

Conceived as an endless modification and adjustment, the deployment of SRM would not be a one-time action with a final completed result but a continuous process of maintenance. Yet to think about SRM as a project of ongoing maintenance changes the ethical evaluation of geoengineering. As Pak-Hang Wong has recently argued, "the requirement of maintenance is of paramount importance to the ethics of geoengineering because it points to a burden and a responsibility that our decision-making on implementation of geoengineering ought to account for." The ethical value of any geoengineering process must be conceived as a long-term process, not a singular event (Wong 2014, 188–90). And if climate maintenance is a never-ending process, it means that humanity is never really in control. There will be a constant need to respond to the demands of the natural cycle. Galarraga and Szerszynski again cite Arendt, who claims that humanity will not be the fabricator of an enduring world but merely "a laboring animal who serves the endless processes of life's self-maintenance" (Galarraga and Szerszynski 2013, 231; citing Arendt 1958).

But for Galarraga and Szerszynski—as well as for me—even the conception or model of SRM as a continuous long-term process is not radical enough to encompass the real meaning of geoengineering. The making of the climate involves the creation of a "radical novelty" or, in other words, "the making of a world" (231). Unlike the models of production and eduction, in which humanity can attempt to mold or to maintain the climate to suit some human end, the process of climate creation is a "historical rupture" with the past, the creation of an entirely new context for human action in the natural world. Galarraga and Szerszynski note that this act of climate creation is even more radical than ecological restoration or remediation, for in these latter processes, an attempt is made to return to some desired past state of the world and the environment. But climate creation by SRM does not and cannot return the world to some past state. In the creation of this new world, new meanings about humanity and nature will arise, and this will require a new analysis of the meaning and content of human responsibility for the planet (231–232).

This analysis is not denied by the proponents of geoengineering. David Keith, perhaps the most well known and credible of the scientists pursuing the geoengineering option as a solution to climate change, has expressed ideas that are in accord with my basic position regarding the ontology and the meaning of human activity in the geoengineering process. In a popular book written about geoengineering, Keith has been quoted as saying,

> With geoengineering, "we [will] gradually engineer a social order that largely is cut off from nature . . . This is why geoengineering is so dangerous . . . It's not the end of nature—but it is the end of wildness—or at least our idea of wildness. It means consciously admitting we're living on a managed planet. It may be that geoengineering can save the Arctic. But it won't be the same Arctic we have today . . . The fact is, whether we want to admit it or not, we're living in a zoo. And we're both the animals and the zookeepers now."
>
> (Goodell 2010, 44–45)

So one of the leading geoengineering scientists admits that the process creates an artifactual reality, a managed reality, where once there existed a free, wild, and autonomous natural system.

In addition, one of the chief philosophical defenders of the project of geoengineering—Christopher Preston—claims that the entire process is anthropocentric in its value judgments and goals. I have discussed Preston's entire argument elsewhere (Katz 2015b) and will not repeat it here. Yet his conclusion embraces anthropocentrism, as he writes, "The significant obligations we have to other humans suggest that even those who value natural processes that have characterized Earth's history might be prepared to interfere with these processes if enough human lives and human suffering were at stake." This means, then, "the anthropocentric position in environmental ethics appears to have won out over the non-anthropocentric one" (Preston 2011, 472).

Geoengineering then only exists as a massive project for the furtherance of human interests. As with the policy of ecological restoration, there is no escape from the fundamental anthropocentrism of the geoengineering project as it attempts to design and control our environment and our world. It will produce a new kind of reality, an artifactual reality, which replaces the natural processes of the Earth. The non-human realm will lose all meaning and value; indeed, its ontological character will change as it is rendered as a mere instrument for the fulfillment of human interests. Thus we return to the concept of "domination." In what many consider the very first philosophical analysis of geoengineering, Dale Jamieson wrote that if geoengineering "were successful, it would still have the bad effect of reinforcing human arrogance and the view that the proper human relationship to nature is one of domination" (Jamieson 1996, 332). I agree with Jamieson, but my conclusion is even stronger: geoengineering requires a fundamental belief in the human domination of the Earth. Once we analyze the meaning, value, and purpose of ecological restoration and geoengineering, we are in the presence of the idea of human omnipotence—unlimited human power—regarding the structure and direction of life. And belief in our omnipotence (and our good intentions) can easily lead to policies and actions that are disastrous in practice and horrible in ethical value. The physical and epistemological domination of all entities and systems of the Earth becomes our goal. This is the promise of the Anthropocene.

IV

Ecological restoration and geoengineering are thus paradigmatic examples of the anthropocentrism of the so-called Age of the Anthropocene. They reveal, in no uncertain terms, the epistemological domination of human thought and meaning imposed on the entire planet. As several authors have recently argued, the very term "Anthropocene" reveals the anthropocentric bias that prevails in thinking about the human relationship to nature. Langdon Winner says that for humans to name the current geologic epoch after ourselves "smacks of an obvious, species-centered narcissism" (Winner 2017, 283). He cleverly undermines the concept by proposing that the era be labeled, instead, the "Langdonpocene" after him! His reasoning—tongue firmly in cheek, we are to suppose—is that in his professional life, he has done more than most people who have ever lived to use up resources and create the environmental crisis of our time. But he is willing to share the honor of naming the current era and is open to other individuals naming this time after themselves in a kind of naming "time-share." Winner's point, of course, is to go beyond this anthropocentric narcissism and acknowledge the presence of the non-human entities of the world (290). Ned Hettinger argues for the same point more directly (without the absurd humor of Winner). Hettinger's analysis shows that the proponents of the Anthropocene concept ignore the spectrum of naturalness that persists on the planet today. It is true that there exists very little of a pristine nature untouched by human activity, but this does not mean that areas relatively free of human influence with thriving non-human populations do not exist and do not have value (Hettinger 2014, 178). Thus a mistaken focus on the Anthropocene as the age in which humans dominate the Earth ignores "the profound role non-human nature continues to play on Earth" (179).

Finally, Eileen Crist's analysis of what she calls the "discourse of the Anthropocene" is the most robust: the Anthropocene label is "a reflection and reinforcement of the anthropocentric actionable worldview" that has generated the environmental crises that confront us (Crist 2013, 130). The discourse of the Anthropocene entails a "Promethean self-portrait" of the human species as being as powerful as Nature (131) but always excluding the non-human from any normative consideration as we humans change our world (133). Crist elaborates that the discourse of the Anthropocene effectively forecloses human choice about alternative means of organizing human life in regard to the non-human natural environment (138, 141). It leaves us with a "project of rationalized domination" (142) as the only acceptable way of approaching the planet. She, instead, argues for a policy (and a philosophical worldview) of what she calls "integration." This would involve rejecting the idea that humans can "take over" or assimilate the natural world into a human-produced planetary system. Integration means rejecting the idea that the Earth is a mere set of resources for the furtherance of human ends. It means rejecting the idea that we humans can manage the Earth. It means that we humans should "pull back and scale down" our impulse to dominate and control the natural environment and instead nurture "a truer vision of Earth as a wild planet overflowing in abundance and creativity" (144).

Let me conclude with a personal anecdote that illustrates the philosophical conclusions developed here. I spend my summers on the barrier beach of Fire Island, off the south shore of Long Island, east of New York City (see Katz 1999, 2002). Fire Island is a designated National Seashore, equivalent to a National Park, although the legislation that created the seashore in the 1960s noted that the island would be of mixed use: residential areas, recreational zones, and wilderness all co-exist. Several days a week, I take a long bike ride along a dirt and gravel road past the historic Fire Island Lighthouse through an area of dunes and beach grass. This area was disrupted by the storm surges of Superstorm Sandy in the autumn of 2012. Sand was pushed inland from the ocean, covering over vegetation and part of the gravel roadway. At one particular bend in the road, an old raised wooden boardwalk that permitted access to the ocean beach was totally destroyed. The Park Service declined to re-build the walkway. Instead, they placed a few yards of snow fence to catch some sand near the dunes and a small sign alongside the road to inhibit any humans from traversing the dunes to get to the beach. The sign states simply, "Area Closed for Dune and Habitat Recovery."

The idea of a "recovery" area for a natural habitat appears to me the exact opposite of the projects of restoration and geoengineering. Here, there is no conscious human intention to design a replacement ecosystem for the one damaged by Sandy. There is, to use Crist's phrase, merely a "pulling back." We erect a small snow fence that itself does minimal damage to the area—and that does not prevent, in reality, any living creature from entering the area—and we do nothing else. There is no plan to bulldoze sand to re-create dunes, no planting of new vegetation, no direct human activity at all. We let the area "recover" from the effects of the storm. We impose no human ideals and no human goals for the area. In doing so, we respect the value of the non-human natural world that exists outside of our manipulation, modification, and control. We reject the human domination of the natural world and the anthropocentrism of the Anthropocene as a solution to the environmental crisis that surrounds us.

Notes

1 Much of the text in this section has been borrowed from my various writings on restoration ecology over the years.
2 Much of the text in this section is derived from Katz (2015b), where the argument is developed in more detail.

References

Arendt, Hannah. *The Human Condition*. Chicago: University of Chicago Press, 1958.
Crist, Eileen. "On the Poverty of Our Nomenclature." *Environmental Humanities* 3 (2013): 129–47.
Drenthen, Martin. "Special Issue: Rewilding in Cultured Layered Landscapes." *Environmental Values* 27, no. 4 (2018).
Galarraga, Maialen and Bronislaw Szerszynski. "Making Climates: Solar Radiation Management and the Ethics of Fabrication." In *Engineering the Climate: The Ethics of Solar Radiation Management*, edited by Christopher Preston, 221–35. Lanham, MD: Lexington Books, 2012.
Goodell, Jeff. *How to Cool the Planet: Geoengineering and the Audacious Quest to Fix Earth's Climate*. Boston: Mariner Books, Houghton Mifflin Harcourt, 2010.
Hettinger, Ned. "Valuing Naturalness in the 'Anthropocene': Now More than Ever." In *Keeping the Wild: Against the Domestication of Earth*, edited by George Wuerthner, Eileen Crist, and Tom Butler, 174–79. Washington, DC: Island Press, 2014.
Jamieson, Dale. "Ethics and Intentional Climate Change." *Climate Change* 33, no. 3 (1996): 323–36.
Katz, Eric. *Anne Frank's Tree: Nature's Confrontation with Technology, Domination, and the Holocaust*. Cambridge, UK: White Horse Press, 2015.
Katz, Eric. "Another Look at Restoration: Technology and Artificial Nature." In *Restoring Nature*, edited by P. Gobster and B. Hall, 37–48. Covelo, CA: Island Press, 2000.
Katz, Eric. "Artefacts and Functions: A Note on the Value of Nature." *Environmental Values* 2, no. 3 (1993): 223–32.

Katz, Eric. "The Big Lie: Human Restoration of Nature." *Research in Philosophy and Technology* 12 (1992): 231–41.

Katz, Eric. "Further Adventures in the Case Against Restoration." *Environmental Ethics* 34, no. 1 (2012): 67–97.

Katz, Eric. "Geoengineering, Restoration, and the Construction of Nature: Oobleck and the Meaning of Solar Radiation Management." *Environmental Ethics* 37, no. 4 (2015): 485–98.

Katz, Eric. "Imperialism and Environmentalism." *Social Theory and Practice* 21, no. 2 (1995): 271–85.

Katz, Eric. "The Liberation of Humanity and Nature." *Environmental Values* 11, no. 4 (2002): 397–405.

Katz, Eric. *Nature as Subject: Human Obligation and Natural Community*. Lanham, MD: Rowman and Littlefield, 1997.

Katz, Eric. "A Pragmatic Reconsideration of Anthropocentrism." *Environmental Ethics* 21, no. 4 (1999): 377–90.

Katz, Eric. "The Problem of Ecological Restoration." *Environmental Ethics* 18, no. 2 (1996): 222–24.

Katz, Eric. "Restoration and Redesign: The Ethical Significance of Human Intervention in Nature." *Restoration & Management Notes* 9, no. 2 (1991): 90–96.

McKibben, Bill. *The End of Nature*. New York, NY: Random House, 1989.

Preston, Christopher. "Re-thinking the Unthinkable: Environmental Ethics and the Presumptive Argument Against Geoengineering." *Environmental Values* 20, no. 4 (2011): 457–79.

Siipi, Helena. "Artefacts and Living Artefacts." *Environmental Values* 12, no. 4 (2003): 413–30.

Siipi, Helena. "Dimensions of Naturalness." *Ethics and the Environment* 13, no. 1 (2008): 71–103.

Tanasescu, Mihnea. "Field Notes on the Meaning of Rewilding." *Ethics, Policy and the Environment* 20, no. 3 (2017): 333–49.

Winner, Langdon. "Rebranding the Anthropocene: A Rectification of Names." *Techne* 21 (2017): 282–94.

Wong, Pak-Hang. "Maintenance Required: The Ethics of Geoengineering and Post-Implementation Scenarios." *Ethics, Policy and Environment* 17, no. 2 (2014): 186–91.

3 Climate ethics bridging animal ethics to overcome climate inaction

An approach from strategic visual communication

Laura Fernández

Introduction

Climate change is thought to be one of the greatest ethical problems of our time and "a severe ethical challenge for humanity and its institutions" (Gardiner 2016, 7). Climate ethics emerged—mainly as an extension to mainstream Anglo-American ethics—as an area of research into the ethical dimensions of climate change, with its central concerns being issues such as "suffering, vulnerability, injustice, rights, and responsibilities" (Gardiner 2016, 49). Climate ethics requires re-conceiving the scope and content of ethics to include "the foreseeable impacts of human activities, and also . . . the foreseeable impacts of policies of nonintervention and inaction" (Attfield 2009, 226). Anthropogenic climate change, however, does not just affect present and future generations of human animals but also affects other non-human beings (Nolt 2011).

As human animals, the hegemonic tendency when researching and acting in the world is to do it from our own senses, from our own bodies and for our own well-being and necessities. Unlike the broad field of environmental and animal ethics, climate ethics has only rarely questioned this anthropocentric and speciesist approach when analyzing the world's realities, especially those related to global climate injustice. In this sense, climate ethics—like many other sets of inter-human ethics—tends to be characterized by an anthropocentric and speciesist bias. This bias omits non-human animals' suffering and interests to live, to avoid suffering, and to be free and does not give moral consideration to the non-human natural world and its intrinsic value (Rolston 1988). Anthropocentric bias in climate ethics limits ethical understandings of the climate change problem.

The aim of this article is to explore the field of visual communication for climate advocacy while using a non-anthropocentric and non-speciesist approach to the global climate change problem. More specifically, I will analyze whether the use of visuals of non-human animal suffering is effective in promoting individual and social attitude-change with respect to climate change.

To this end, this chapter will first address the problem of anthropocentrism and an anthropocentric and speciesist bias in climate ethics. Second, animal ethics will be considered as a tool to contributing to developing a less anthropocentric and speciesist form of climate ethics and to finding the common ground (Kemmerer 2015) between animal and climate advocacy. To exemplify anthropocentric and speciesist bias in climate ethics, the paradigmatic case of the animal agriculture and aquaculture industries will then be analyzed, both because of their huge contribution to climate change and also because of the omission and lack of visibility of non-human animal suffering (both direct and indirect) in the considerations of climate ethics. Finally, in order to construct a non-human animals-inclusive form of climate ethics and advocacy, an analysis from climate and animal advocacy visual campaigns will be presented to examine whether visuals of non-human animal suffering are effective in promoting attitude change in favor of the moral consideration of non-human animals and against climate change.

These considerations show that climate ethics and advocacy can benefit from the strategic use of images of non-human animal suffering because of their potential to create an emotional appeal to the global and frequently abstract issue of climate change. This emotional appeal is considered vital to moralizing attitudes and to motivating mitigating actions. Therefore, in this chapter, I argue that the inclusion of visuals of non-human animal suffering is both an ethical and strategically effective approach for developing a less anthropocentric climate ethics and advocacy.

The environmental ethics tradition, speciesism, and the anthropocentric bias in climate ethics

Unlike anthropocentric inter-human ethics, from the very start, environmental ethics was born from the "desire to formulate ethical theories that did a better job of accounting for our moral obligations to the non-human natural world" (McShane 2009, 407). Under this premise, several debates have taken place, creating distant positions in what can globally be considered "environmental ethics." Regarding the intrinsic value and moral consideration given to non-human animals, two main traditions can be discerned: individualism— which includes biocentrism and animal rights—and holism, or ecocentrism. The individualist positions in environmental ethics support the idea that what matters morally are individual beings (human or non-human). The main criterion for moral consideration in animal utilitarian ethics is sentience. Utilitarians (Singer 1990) define sentience as the capacity to suffer and enjoy, which means non-human animals can be harmed or benefited. In Reagan's deontological ethics, non-human animals have the status of "subjects-of a life" (1983), and in the case of ecofeminism and the ethics of care, non-human animals are subjects of human care and compassion (Donovan and Adams 2007). In contrast, the ecocentric/holistic paradigm in environmental ethics focuses on whole entities—such as species, ecosystems, or the biosphere—as well as their interdependence, and generally considers individuals in terms of their contributions to these larger entities.[1]

Ecocentrism is undoubtedly a non-anthropocentric environmental ethics; however, in practice, it frequently fails to recognize the value of individual living beings. In so doing, ecocentrism has a speciesist bias and is, in this sense, ethically incompatible with non-speciesist approaches (Faria 2012). As Catia Faria (2012) argued, environmental interventions for restoring an ecosystem's balance would allow for only one exception: when human lives are at risk in those interventions. In this sense, in efforts to recover the balance of ecosystems, violations of human rights, such as the right to live and not to be harmed, would not be considered acceptable, and yet, such violations are commonly and systematically perpetrated against non-human animal individuals from species considered to be "invasive." In this chapter, it is argued that a non-anthropocentric ethical approach must morally consider all living beings, no matter what their species or their instrumental value for human interests.

A discussion of the individualist and holist positions within the environmental ethics debate is beyond the scope of this chapter, which is focused on visual communication for climate advocacy. It is important to clarify, however, that this essay is based on the broad work done in the area of environmental and animal ethics and calls for a biocentric and non-speciesist approach to climate ethics that recognizes speciesism as an oppressive ideology (Nibert 2002) that needs to be rejected and advocates for the flourishing of all living beings. As Rebekah Humphreys declares in her chapter in this anthology: "All nonsentient creatures have interests in thriving after their own kind, exercising their species-specific tendencies, and fulfilling potentialities" (2020, 49).

For this, as well as practical reasons, in this chapter, I will focus on speciesism within climate ethics and on the visual representation of non-human animal suffering. In this sense, I follow Attfield's argument when he claims that "biocentrists should be open to systemic thinking, but do not need to supplement their understanding of moral standing beyond the individual creatures of the present and the future" (Attfield 2020, 63).

Speciesism has been defined in a moral sense as "the unjustified disadvantageous consideration or treatment of those who are not classified as belonging to a certain species" (Horta 2010, 1). Speciesism (Ryder 1970/2000; Singer 1990) has also been considered in a structural sense as a system of domination and oppression (Nibert 2002) that overlaps other power structures and systems of power such as colonialism (Belcourt 2015), hetero-patriarchy (Adams 1990/2010; Gaard 2017), Earth domination (Kemmerer 2015), ableism (Taylor 2014), and capitalism (Nibert 2002; Hribal 2010), among others.

Following these overlaps of oppression, ecofeminists, animal ethicists, and critical animal studies scholars have sought to reveal the anthropocentric and speciesist bias in ethics, politics, and other fields of knowledge. The anthropocentric bias is parallel to the Eurocentric and androcentric bias denounced by feminist, gender, de-colonialist, and critical race scholars and shows the binarist roots of anthropocentric, hetero-patriarchal, white, Western thought. The challenge here is to revise the bias in the history of knowledge by adopting what Best (2014) calls the "animal standpoint," instead of continuing with the tendency of reproducing this bias against those who are outsiders of a certain group in the conception of ethics (Nolt 2011, 705).

Regarding climate ethics, the anthropocentric bias provokes a partial analysis of climate change, placing human animals in the center of moral consideration and privileging them and their well-being over those of other animals for speciesist reasons. This bias enormously affects our understanding of the effects of climate change on non-human animals in both the short and long term and is a limitation for both adaptation and mitigation policies (Nolt 2011).

The focus of this chapter will now turn to non-human animals and the anthropocentric and speciesist bias in climate ethics for practical reasons but on the basis of the idea that a non-anthropocentric approach to climate ethics must include a critical assessment of our relations with other animals and the natural non-human world. This idea is more broadly argued throughout the present anthology, especially in the work of Attfield, Nolt, Rolston, and Humphrey.

How can climate ethics deal with anthropocentric bias? Contributions from animal ethics

As Nolt remarks, "the fundamental assumption of non-anthropocentric value theory is that some non-human entities can be benefited or harmed, independently of what we think, value, or desire" (2011, 702). Following this stance as applied at least to non-human animals, human animals should have extended responsibilities towards non-humans not only for directly caused suffering but also the present and potential consequences of climate change, which is mainly anthropogenic.[2] Nolt (2011) presents an integrated value ethical theory from environmental, animal, and intergenerational ethics in which he argues for a long-term non-anthropocentric climate ethics. He explains that climate ethics and policy need to be long-term, because "benefit and harm are not confined in the present or near present" (2011, 103), and these harms exist independently of human animals; they existed before we did and will continue to do so after we become extinct.

The climate crisis affects both human and non-human animals. In many cases, the desirable ethics and politics for combating the climate change problem may converge for both human and non-human animals. The idea that both anthropocentric and non-anthropocentric approaches to the issue would recommend the same environmental policies is known as Brian Norton's "convergence hypothesis." However, this hypothesis has been questioned from the non-anthropocentric perspective (Stenmark 2002; Nolt 2011; McShane 2016). It is not always granted that advancing the interests of human animals would mean an advance in the interests of non-human animals. On the contrary, it looks as if, in some cases, these interests may differ and even clash. McShane outlines, for example, the potential conflicts over land use and other limited access to what makes life possible (2011, 197). An apparently beneficial policy for humans that is not specifically designed to consider non-human interests will probably harm non-humans. In this regard, Humphreys (2020) notes that

> even in scenarios in which stricter mitigation policies were adopted than those proposed by the Paris Agreement, the interests of most of the world's living

creatures (nonsentient beings), present and future ones included, would fail to be included as beings that should be given direct consideration with regards to appropriate policies and practices.

(p. 60).

Considering the anthropocentrism in climate ethics, I suggest here that animal ethics can contribute to critically revising the anthropocentric bias present in mainstream climate ethics. Since the very beginning, animal ethics has argued for the consideration of other animals. The attribution of moral consideration to non-human animals is an indispensable step to including them and their interests in climate ethics and politics. The main premise of egalitarianism in animal ethics is that improving "the situation of the worse off is more important than benefiting the ones who are better off" (Horta 2013, 8). Unlike the convergence hypothesis, the priority perspective in animal ethics considers that what is valuable in interspecies relations "is not how equally value is distributed among individuals but rather how individual well-being stands in absolute terms" (Faria 2014, 5). Animal rights ethicist Tom Regan (1983) considered not only the final amount of suffering or welfare—as the utilitarian perspective usually does (Singer 1990)—but also the intrinsic value of each individual animal, considering them "subjects of a life."

If we accept that non-human animals deserve our moral consideration and that speciesism should be rejected, a clear non-anthropocentric approach to climate ethics and policies is needed. McShane's proposals along these lines (2016)—to conduct research on the impact of current international policies on non-human animals, to assess how climate change policies apply to non-human animals and interspecies justice, and to reject the use of a standard economic methodology when approaching the suffering and death of other animals—embody such an approach. Animal ethics would also reject non-anthropocentric speciesism[3] and advocate for broadening the circle of compassion by not morally distinguishing non-human animals by their species nor by the wild/domesticated binary.

The next section will show how an anthropocentric bias is inherent in climate ethics by the paradigmatic example of the animal agriculture and aquaculture industries. I will also present proposals for changing this approach by looking for common ground between climate advocacy and animal advocacy.

Animal agriculture and aquaculture industries and climate advocacy: where are the non-human animals and their suffering?

In many analyses of the global problem of climate change, it has been widely argued that the food system in general, and the animal agriculture and aquaculture industries in particular, play a decisive role in current global warming trends (Goodland and Anhang 2009; Henning 2011; Leip et al. 2010; Scarborough et al. 2014; Steinfeld 2006; Worldwatch Institute 2004). Some of the main problems these industries contribute to include water pollution and waste, soil degradation and deforestation, energy waste, bio-contamination and diseases, and species extinction, as well as greenhouse gas emissions and air pollution. Despite agri-business's position as a primary producer of climate change, generating from 18 (Steinfeld 2006) to 51 percent (Goodland and Anhang 2019) of the global green-house gas emissions (GHG), environmental and climate advocacy have largely ignored this fact, and the majority of advocacy organizations have not defended dietary choice as a climate change mitigation strategy until only very recently.

I will now turn to examine several studies on climate advocacy in order to emphasize the links between climate ethics (theory) and climate advocacy (prac-tice), understanding that they are complementary and need each other. In the exercise of putting into practice climate ethics as climate policies and climate edu-cation, the anthropocentric and speciesist theoretical bias in the advocacy arena becomes abundantly clear.

Firstly, the environmental and climate advocacy movements have not paid enough attention to the animal agriculture and aquaculture industries and their terrible impacts on other animals and the environment. This omission, I will argue, can be explained by the psychological mechanisms that bring humans to avoid recognizing non-human animals as moral subjects, by the political economy of animal exploitation and the media's role in manufacturing consent in society about speciesism (Almiron, Cole, and Freeman 2016) and climate change denial, and by the previously explained anthropocentric bias in climate ethics, which does not consider non-human animal well-being as morally significant.

Laestadius, Neff, Barry, and Frattaroli (2014) interviewed NGO staff members from environmental, animal protection, and food-focused NGOs in the United States, Sweden, and Canada and asked them about the active meat-reduction campaigns of their organizations (2014, 33). Their research showed that only a small number of the NGOs linked meat consumption to climate change and that they were even less likely to dedicate a specific campaign to public education on the issue. The NGOs' discourse pointed out different factors when adopting campaigns:

> (1) whether the issue fits in with the NGO's core missions; (2) whether the issue fits in with the NGO's tactical preferences; (3) the perceived outcome of engagement with the issue; and (4) the NGO's capacity to take on the issue.
>
> (2014, 35)

As Cole et al. (2009, 166) concluded in their research on the role of animal agri-culture and aquaculture in consumer behavior and personal awareness regarding climate change, "There is a generalized lack of awareness about the relationship between animal farming and climate change."

The reasons for this general lack of awareness about the connection between non-human animal exploitation and climate change are both psychological and structural. Research in the field of social psychology has shown that human ani-mals construct psychological defenses when there is an incompatibility between their values and attitudes. This psychological defense is called "cognitive dissonance" (Joy 2010) and works by compartmentalizing our feelings towards non-human animals, thereby allowing us to deny our responsibility and remain morally indifferent towards their exploitation and slaughter. Humans' general lack of awareness can also be analyzed from a political-economy perspective. Agribusiness lobbies use economic and political power to hide their practices and generate manufactured consent in society with the help of the media and academia, frequently in the form of think-tanks and discourse coalitions (Almiron 2017). Animal industries broadcast the message that consuming animal products is necessary, normal, and natural—Joy's (2010) 3Ns of justification for carnism[4]—and that this practice is completely unrelated to climate change.

The animal agriculture and aquaculture industries have long denied the reality of farmed animal[5] suffering and hardly ever taken into account the indirect suffering of wild animals. Gaard (2017) would call the denial of granting moral consideration to individual wild animals "environmental speciesism," for such animals are only considered morally if they are a member of a species in danger of extinction. Along these lines, stopping the consumption of non-human animals and their by-products and adopting a vegan diet would not only be the most sustainable option and an effective strategy for mitigating climate change (Scarborough et al. 2014) but also a non-anthropocentric ethical position against non-human animal oppression and domination.

As mentioned before, speciesism is also linked with other forms of oppression. In 1993, Fox already stated in his paper "Environmental Ethics and the Ideology of Meat Eating":

> Meat is the quintessential symbol of our species' domination of nature, our capacity to transform life into death, to conquer and exploit what is other, what is at our mercy. Meat is also a highly visible symbol, reminder and reinforcer of patriarchal control in all of its manifestations. Meat is masculine food, powerful food; to be a "real man" is to eat meat, lots of it, and the redder the better.
>
> (p. 128)

Besides the hegemonic tendency of reproducing human supremacy and distorted representations of non-human animals and their relations—what scholars such as Nibert (2002) and Khazaal and Almiron (2016) call "speciesist ideology"—we have evidence that supports the importance of considering non-human animal suffering and as well as representing it not just because of its ethical nature but also because of the potential of its emotional impact to make it an effective visual strategy against climate change in animal and climate advocacy campaigns.

Non-human animal-inclusive climate advocacy: the importance of visual campaigns

Climate and animal advocacy have frequently had different priorities: the adoption of mitigation and adaptation strategies to fight climate change in the case of the former and the promotion of well-being and moral consideration of non-human animals in the case of the latter. Nevertheless, following Kemmerer, environmental—and climate is here included inside the big umbrella of environmental ethics and advocacy—and animal advocates have much in common. They share an interest in expanding our moral circle and "the desire to eradicate anthropomorphism/speciesism/humanocentrism" (2015, 6). This shared goal constitutes what Kemmerer (2015) calls the "common ground." In this section, I will consider some research from the field of visual communication on the effectiveness of using images of non-human animals suffering in animal and climate advocacy, while also looking for the common ground shared by non-human animal-inclusive climate ethics and advocacy.

We live in highly visual societies, in which we are constantly exposed to a large amount of visual information by the mainstream and social media and advertisements. As Russmann and Svensson observe,

> the sharing of images is becoming an integral part of the social media experience today and given that social media platforms are the prime locus for sociability—at least among young people in the West—this shift towards visuals arguably transforms how we relate to each other and the world around us, as well as how we perceive and construct our sense of self.
>
> (2017, 1)

In this context, climate and animal advocates use visuals as an essential tool to create awareness and promote attitude change towards problems such as climate change and animal suffering and to counteract those representations that deny climate change or distort the realities of non-human animals' suffering.

When analyzing the impact of an image about climate change, for instance in the media, effective visual communication researchers take into account two main variables: salience and efficacy. Salience refers to whether the issue presented by the image (in this case, climate change) is considered relevant. Efficacy, in contrast, appeals to the individual's agency regarding the exposed reality, that is, the sense that they are able to do something about climate change (O'Neill 2013). While salience is more related to the mainstream reception of the issue, efficacy refers more to the individual's response to a global issue. The strategic promotion of one or the other variable depends on the audience. Finding balance between dimensions is important, because when an image has great salience, it tends to create a backlash in personal action, while when the image promotes efficacy, salience is put aside. It is difficult to find both characteristics in the same image (O'Neill et al. 2013).

The research on the effectiveness of the use of images of non-human animal suffering in causing attitude change shows that such images do provoke an emotional appeal, and this appeal easily leads audiences to moralize their attitudes or to adopt attitude change (Wisneski and Skitka 2017). The existing research, however, is not very abundant, and more research needs to be done on this topic. For example, according to Rose's structural analysis (2016), all the different levels of meanings that make up an image should be examined: the sites of the image (the image itself), the sites of production (where it was made), the sites of circulation (where it was spread), and the audiencing (interaction with the audience). I will not focus on the four levels in this chapter, but different strategies related to research on the different sites will be explored in more detail.

Persuasive, effective communication bases its knowledge on emotions and their role for social change. Images of non-human animals have generally been framed as emotional appeals to individual and collective attitude change. Much of the social psychology field agrees on the idea that there is an entanglement between emotion and morality (Swim and Bloodhart 2015; Wisneski and Skitka 2017). As Jasper observes, "emotions, freed from the pejorative mind-body dualisms of the past, promise to advance our comprehension of agents and their motivation" (2011, 298). Emotions play a central role in social movements and activism, pervade all social life, and are behind our actions and convictions: they can engage us with a social cause and also are an essential factor for long-term commitment and the creation and maintenance of activist communities (Jasper 2011).

With the goal of discerning the effectiveness of the use of visuals of non-human animal suffering in promoting individual and social attitude-change regarding a non-anthropocentric climate advocacy in mind, we may ask ourselves: How can suffering be strategically framed to promote awareness and mobilization among audiences? Swim and Bloodhart (2015) researched the effectiveness of images of injured polar bears with both an empathic and objective perspective to lead audiences to engage with climate advocacy. They obtained very useful conclusions (2015, 465): (1) The images evoked emotions both with and without perspective-taking instructions, so the message may be helpful, especially with non-environmentalists who are not exposed to these visuals; (2) "it may be necessary to encourage audiences to take an empathic perspective in order to move individuals from focusing on their own feelings of, for instance, distress or sadness, to more complex other-focused feelings in order to engage climate change activism" (465); (3) emotions such as hope and empathy were important mechanisms to support climate change advocacy; (4) regarding negative emotions, "a decrease in boredom and an increase of worry and personal guilt" (464) can be potentially effective in leading audiences to engage in climate change activism, especially when the audience is non-environmentalist.

Some years earlier, Huddy and Gunnthorsdottir (2000) conducted research on the relevance of the animal species and the "cuteness" or "ugliness" of the animal to generate positive feelings and engagement with their environmental conservation. Even though they were not examining images of non-human animal suffering, but rather of extinction, their main conclusions are useful for the goal of this chapter. They found that mammals were considered more positively than insects and "cute" mammals—for example, a monkey as opposed to a bat—were considered the most effective images to generate engagement, whereas the image of an "ugly" insect—a bug instead of a butterfly—was found to generate less engagement with the conservation goal. It is necessary to point out that the latent idea underlying these results is that humans are more able to emotionally connect with non-human animals that are similar to them and which they consider beautiful. This is an anthropocentric reality that provides some strategic direction but, at the same time, must be questioned; it shows that measures need to be taken not to forget about other non-human animals who are just as exploited and affected by climate change, despite not being mammals or as photogenic. The inclusion of this critique of our anthropocentric gaze and imaginary must be a fundamental task for a non-anthropocentric climate ethics and advocacy, even if it means not being fully strategic at first.

As we have seen, visuals depicting suffering of animals who live in nature have been effective in raising awareness about climate change and its effects. As will be shown, we have strong arguments to believe that the same thing would happen with farmed animals. To approach a non-anthropocentric climate advocacy, the proposal here is to combine visuals of both wild and domesticated animals. This premise is based on both ethical and strategical arguments. Many wild animals are directly and indirectly affected by climate change, a clear example being those affected by the sixth mass extinction event currently underway in the Anthropocene (Crutzen 2006), as is also mentioned by Nolt in this anthology (2020). If climate visual advocacy also includes the suffering of farmed animals in representations of climate change, it will contribute to shedding light on the terrible implications of the animal agriculture and aquaculture industries.

Representing non-human animals is an effective way of promoting both salience (with the idea that the suffering is real and present) and efficacy (while suggesting mitigation measures such as a vegan dietary choice). In the following, some animal advocacy campaigns and strategies will be suggested to represent climate change through the use of visuals of non-human animal suffering. This strategy also promotes the recognition of common ground between climate and animal advocacy, by following the non-speciesist goal of breaking the species barrier and by nourishing climate advocacy from what we know about strategic visual communication for animal advocacy.

The previous differentiation between species and the body characteristics of non-human animals may be a consequence of non-anthropocentric speciesism and the attribution of moral consideration to those animals that are more similar to us or that make us feel good. To break with our speciesist gaze toward other animals and consider them as equals, some animal scholars and advocates have defended the moral shock frame of animal suffering, which is characterized by its explicit representation of violence and which has been generally applied to document farmed animals' realities in factory farms and slaughterhouses.

In 1995, Jasper and Poulsen coined the concept of "moral shock" and described it as a stimulus that causes a sense of outrage, which, in turn, leads individuals to react in response to it. Moral shock strategies have been one of the main strategies for "recruitment" of people with no prior relationship with animal advocacy (Jasper and Poulsen 1995; Jasper and Nelkin 2007). Wisneski and Skitka (2017) confirm Jasper's hypothesis of the effectiveness of moral shock for attitude change in their research. They consider moral shock as unique to morality: "attitudinally relevant, disgust inducing stimuli led to increased moral conviction, but did not affect attitude importance or extremity. Moral shocks, therefore, appear to moralize attitudes without affecting other dimensions of attitude strength" (Wisneski and Skitka 2017, 147). Although moral shock has been defended and supported by a number of scholars and researchers (Jasper and Nelkin 2007; Scudder and Bishop-Mills 2009; Wisneski and Skitka 2017, among others), its use is still a controversial topic within the field of strategic communication because, as it is linked with emotions, it also has the risk of generating rejection and offense and even of promoting backlash or demobilization among audiences (Mika 2006; Sullivan and Longnecker 2010).

Some authors have shown their fears regarding the ethical representation of non-human animals in the shock frame, considering the problems of compassion fatigue—defined as "the wearing out of the ability to care about suffering" (Aaltola 2014, 28)—among activists and the privacy of the non-human animals represented. As philosopher Elisa Aaltola notes, the privacy of non-human animals is violated "when their subjectivity and personhood are ignored and replaced with empty forms or caricatures that invite nothing but aesthetic amusement" (2014: 26). Atlas (2014) shared this perception and argued the importance of communicating the victim's personhood. After all, "moral shocks require the activation of attitudinally relevant disgust and conscious awareness of the source of that disgust" (Wisneski and Skitka 2017, 147–48). To counter these ethical risks, the proposal of Aaltola (2014) and Atlas (2014) is to approach and/or complement moral shock with an agency perspective, wherein non-human animals are represented not as objects but as subjects. Therefore, other frames and communication strategies could be combined with moral shocks and empathic approaches in order to promote efficacy and avoid audience backlash. Sullivan and Longnecker (2010) presented two other frames for non-human animal representation: those of "animal welfare as a social norm" and "non-human animals as intelligent beings." However, these frames do not challenge anthropocentrism and, therefore, are not sufficiently effective for non-anthropocentric climate advocacy. As argued by communication scholar Carrie P. Freeman (2009), "authenticity" between the ideology and the frame of a message is an essential value for advocacy communication. Some relevant non-anthropocentric frames to represent animals are those that depict rescued animals—such as visuals of non-human animals in animal sanctuaries that let us "imagine how the absence of fear and suffering might feel" (Cronin and Kramer 2018, 90)—or those which highlight non-human animal agency (Hribal 2010) and the dignity of the represented non-human subjects (Atlas 2014; Aaltola 2014) instead of their beauty or photogenicity (as in Huddy and Gunnthorsdottir 2000).

In parallel, Elisabeth Cherry pointed out boundary work as "both a strategy and a goal for new social movements that seek cultural change" (2010, 452). Cherry holds that symbolic boundaries maintain social differences between groups and legitimate the inferiorization of one group in front of the other. In the case of non-human animals, symbolic boundaries morally organize, within a hierarchy, which animals deserve moral consideration and which ones do not. These boundaries express themselves in the human/non-human animal boundary or in the companion/farmed animal boundary. To break this barrier between species, Cherry proposes boundary shifting. She notes two main strategies: (1) boundary blurring, which focuses on universalizing victimization, for example by talking about meat as murder or universalizing struggles by focusing on animal rights as an aim of social justice and (2) boundary crossing, where "activists engage physically, discursively and iconographically" (p. 468), for example by substituting a companion animal body for a farmed animal body.

As has been shown, the suffering frame in visual campaigns, varied in its multiple forms, may help both climate and animal advocates by raising awareness about the huge, transversal issue of climate change. Promoting a visual discourse based on common ground can motivate connections between struggles (such as climate and animal advocacy) and also brings a global understanding of the problem of climate change by connecting the animal agriculture and aquaculture industries with their effects on animal suffering: both of exploited animals and of those which are living in the wild but are and will be affected by the bio-contamination of water, deforestation, desertification, water waste, and GHG emissions, among others.

Conclusions: bridges to overcome the global climate crisis

In this chapter, I have argued that a long-term non-anthropocentric climate ethics that locates non-human animals in the center of moral consideration and policy design is necessary to stop reproducing the speciesist oppression of other animals. For this goal, animal ethics contributes to the construction of a non-anthropocentric climate ethics that challenges the speciesist system of domination.

The anthropocentric bias in traditional climate advocacy has been addressed using the paradigmatic cases of the animal agriculture and aquaculture industries and the way they are insufficiently targeted by the climate advocacy movements. A special emphasis has been made on the invisibility of the suffering of wild and farmed animals, who are directly or indirectly affected by this industrial complex. A vegan diet has been suggested as an effective way of mitigating climate change and as a non-speciesist approach to the issue.

In particular, it has been argued that representing the problem of climate change by using visuals of animal suffering can be effective for promoting individual and social attitude change in the direction of a non-anthropocentric approach to climate change. Strategic visual communications that represent non-human animal suffering offer both climate and animal advocacy groups a variety of possibilities to engage audiences through emotional appeals. Emotional appeals make possible concrete connections with the global and frequently abstract issue of climate change and promote ethical attitude change leading to greater commitments toward mitigating policies. At the same time, this inclusive approach to visual communications motivates the quest for common ground between climate and animal advocates and their connected struggles for non-anthropocentric climate advocacy.

Acknowledgments

Research for this chapter has been conducted with the support of Generalitat de Catalunya (Department of Universities and Research–AGAUR) and the European Social Fund.

Notes

1 The consideration of individuals in ecocentric/holistic environmental ethics is mainly instrumental, but there are exceptions such as Rolston (1988), who also takes into consideration the intrinsic value of the individual organism.
2 However, from an antispeciesist approach, human animals may consider ethically not only the anthropogenic-based climate consequences on non-human animals but also those non-anthropogenic causes that threaten non-human animals, such as diseases, starvation, dehydration, or weather conditions. In these cases, positive interventions in nature may be necessary to help non-human animals in need. More information on this debate can be found in C. Faria and E. Paez, "Animals in need. The problem of wild animal suffering and intervention in nature," *Relations 3*, no. 1 (2015).
3 Non-anthropocentric speciesism has been defined as the unjustified priority given to the interests of some non-humans (e.g. dogs) over the interests of other non-humans (e.g. pigs), based on species-membership, following C. Faria, and E. Paez, Anthropocentrism and speciesism: Conceptual and normative issues. *Revista de bioética y derecho 32* (2014): 82–90.
4 Carnism is the invisible belief system, or ideology, that conditions people to eat certain animals. Carnism is essentially the opposite of veganism, as "carn" means "flesh" or "of the flesh" and "ism" refers to a belief system (Joy 2010).
5 Following Freeman (2014), I have decided to use the term *farmed* animal instead of *farm* animal to underline the idea that "farming is something we do to these individuals—something we force upon them" (29). Reference: C. P. Freeman, *Framing farming: Communication strategies for animal rights* (New York: Rodopi, 2014).

References

Aaltola, E. "Animal Suffering: Representations and the Act of Looking." *Anthrozoös* 27, no. 1 (2014): 19–31. doi:10.2752/175303714X13837396326297.

Adams, C.J. *The Sexual Politics of Meat: A Feminist-Vegetarian Critical Theory*. New York: Continuum, 1990/2010.

Almiron, N. "Favoring the Elites: Think-Tanks and Discourse Coalitions." *International Journal of Communication* 11 (2017): 4350–69.

Almiron, N., M. Cole, and C.P. Freeman. *Critical Animal and Media Studies: Communication for Non-Human Animal Advocacy*. New York: Routledge, 2016.

Atlas, K. "Allies and Images: The Importance of Communicating the Victim's Personhood." *The Liberationist*. 2014, October 28. www.directactioneverywhere.com/theliberationist/2014/10/28/allies-and-images-the-importance-of-communicating-the-victims-personhood

Attfield, R. "Biocentrism, Climate Change and the Spatial and Temporal Scope of Ethics." In *Climate Change Ethics and the Non-Human World*, edited by B.G. Henning and Z. Walsh, 63–74. 2020.

Attfield, R. "Mediated Responsibilities, Global Warming, and the Scope of Ethics." *Journal of Social Philosophy* 40, no. 2 (2009): 225–36.

Belcourt, B-R. "Animal Bodies, Colonial Subjects: (Re)locating Animality in Decolonial Thought." *Societies* 5, no. 1 (2015): 1–11. doi:10.3390/soc5010001.

Best, S. *The Politics of Total Liberation*. New York: Palgrave Macmillan, 2014.

Cherry, E. "Shifting Symbolic Boundaries: Cultural Strategies of the Animal Rights Movement." *Sociological Forum* 25, no. 3 (2010): 450–75.

Cole, M., M. Miele, P. Hines, K. Zokaei, B. Evans, and J. Beale. "Animal Foods and Climate Change: Shadowing Eating Practices." *International Journal of Consumer Studies* 33, no. 2 (2009): 162–67. doi:10.1111/j.1470–6431.2009.00751.x.

Cronin, J.K. and L.A. Kramer. "Challenging the Iconography of Oppression in Marketing: Confronting Speciesism Through Art and Visual Culture." *Journal of Animal Ethics* 8, no. 1 (2018): 80–92.

Crutzen, P.J. "The 'Anthropocene'." In *Earth System Science in the Anthropocene*, edited by E. Ehlers and T. Kraft, 13–18. Berlin: Springer, 2006.

de Boer, J., H. Schösler, and J.J. Boersema. "Climate Change and Meat Eating: An Inconvenient Couple?" *Journal of Environmental Psychology* 33 (2013): 1–8. doi:10.1016/j.jenvp.2012.09.001.

Donovan, J. and C.J. Adams, eds. *The Feminist Care Tradition in Animal Ethics*. New York: Columbia University Press, 2007.

Faria, C. "Equality, Priority and Non-Human Animals." *Dilemata* 14 (2014): 225–36.

Faria, C. "Muerte entre las flores: el conflicto entre el ecologismo y la defensa de los animales no humanos." *Viento Sur* 125 (November 2012): 67–76.

Fox, M.A. "Environmental Ethics and the Ideology of Meat Eating." *Between the Species* 9, no. 3 (1993): 121–32. doi:10.15368/bts.1993v9n3.2.

Freeman, C.P. "A Greater Means to the Greater Good: Ethical Guidelines to Meet Social Movement Organization Advocacy Challenges." *The Journal of Mass Media Ethics* 24, no. 4 (2009): 269–88. doi:10.1080/08900520903320969.

Gaard, G. "Feminism and Environmental Justice." In *Handbook of Environmental Justice*, edited by R. Holifield, J. Chakraborty, and G. Walker, 74–88. New York, NY: Routledge, 2017.

Gardiner, S.M. "In Defense of Climate Ethics." In *Debating Climate Ethics*, edited by S.M. Gardiner and D.A. Weisbach. Oxford Scholarship Online, 2016. www.oxfordscholarship.com.sare.upf.edu/view/10.1093/acprof:oso/9780199996476.001.0001/acprof-9780199996476

Goodland, R. and J. Anhang. "Livestock and Climate Change. What If the Key Actors in Climate Change Are Cows, Pigs and Chickens?" *World Watch Magazine* 22, no. 6 (November/December 2009): 10–19. www.worldwatch.org/node/6294.

Henning, B.G. "Standing in Livestock's 'Long Shadow': The Ethics of Eating Meat on a Small Planet." *Ethics and the Environment* 16, no. 2 (2011): 63–93. doi:10.2979/ethicsenviro.16.2.63.

Horta, O. "The Moral Status of Animals." In *The International Encyclopedia of Ethics*, edited by H. Laffoyette, J. Deigh, and S. Stroud. Wiley Online Library, 2013. doi:10.1002/9781444367072.wbiee156.

Horta, O. "What Is Speciesism?" *The Journal of Agricultural and Environmental Ethics* 23, no. 3 (2010): 243–66. doi:10.1007/s10806-009-9205-2.

Hribal, J. *Fear of the Animal Planet: The Hidden History of Animal Resistance*. Oakland, CA: AK Press, 2010.

Huddy, L. and A.H. Gunnthorsdottir. "The Persuasive Effects of Emotive Visual Imagery: Superficial Manipulation or the Product of Passionate Reason?" *Political Psychology* 21 (2000): 745–78.

Humphreys, R. "Suffering, Sentientism and Sustainability: An Analysis of a Non-Anthropocentric Moral Framework for Climate Ethics." In *Climate Change Ethics and the Non-Human World*, edited by B.G. Henning and Z. Walsh, 49–62. 2020.

Jasper, J.M. "Emotions and Social Movements: Twenty Years of Theory and Research." *Annual Review of Sociology* 37 (2011): 285–303. doi:10.1146/annurev-soc-081309-150015.

Jasper, J.M. and D. Nelkin. "The Animal Rights Crusade." In *Life in Society: Readings to Accompany Sociology, a Down-to-Earth Approach*, edited by J.M. Henslin, 225–32. Edwardsville, IL: Southern Illinois University and Pearson, 2007.

Jasper, J.M. and J.D. Poulsen. "Recruiting Strangers and Friends: Moral Shocks and Social Networks in Animal Rights and Anti-Nuclear Protests." *Social Problems* 42, no. 4 (1995): 493–512.

Joy, M. *Why We Love Dogs, Eat Pigs, and Wear Cows*. San Francisco, CA: Conari Press, 2010.

Kemmerer, L., ed. *Animals and the Environment*. New York: Routledge, 2015.

Khazaal, N. and N. Almiron. "An Angry Cow Is Not a Good Eating Experience." *Journalism Studies* 17, no. 3 (2016): 374–91. doi:10.1080/1461670X.2014.982966.

Laestadius, L.I., R.A. Neff, C.L. Barry, and S. Frattaroli. "'We Don't Tell People What to Do': An Examination of the Factors Influencing NGO Decisions to Campaign for Reduced Meat Consumption in Light of Climate Change." *Global Environmental Change* 29 (2014): 32–40. doi:10.1016/j.gloenvcha.2014.08.001.

Leip, A., F. Weiss, T. Wassenaar, I. Perez, T. Fellmann, P. Loudjani, and K. Biala. *Evaluation of the Livestock Sector's Contribution to the EU Greenhouse Gas Emissions (GGELS)—Final Report*. European Commission, Joint Research Centre, 2010. https://ec.europa.eu/agriculture/external-studies/livestock-gas_en.

McShane, K. "Anthropocentrism in Climate Ethics and Policy." *Midwest Studies in Philosophy* 40, no. 1 (2016): 189–204. doi:10.1111/misp.12055.

McShane, K. "Environmental Ethics: An Overview." *Philosophy Compass* 4, no. 3 (2009): 407–20. doi:10.1111/j.1747-9991.2009.00206.x.

Mika, N. "Framing the Issue: Religion, Secular Ethics and the Case of Animal Rights Mobilization." *Social Forces* 85, no. 2 (2006): 915–41.

Nibert, D.A. *Animal Rights, Human Rights*. Lanham, MD: Rowman & Littlefield Publishers, 2002.

Nolt, J. "Climate Change and the Loss of Non-Human Welfare." In *Climate Change Ethics and the Non-Human World*, edited by B.G. Henning and Z. Walsh, 10–22. 2020.

Nolt, J. "Nonanthropocentric Climate Ethics." *WIREs Climate Change* 2, no. 5 (2011): 687–700. doi:10.1002/wcc.132.

O'Neill, S.J. "Image Matters: Climate Change Imagery in US, UK and Australian Newspapers." *Geoforum* 49 (2013): 10–19. doi:10.1016/j.geoforum.2013.04.030.

O'Neill, S.J., M. Boykoff, S. Niemeyer, and S.A. Day. "On the Use of Imagery for Climate Change Engagement." *Global Environmental Change* 23, no. 2 (2013): 412–21. doi:10.1016/j.gloenvcha.2012.11.006.

Regan, T. *The Case for Animal Rights*. Berkeley, CA: University of California Press, 1983.

Rolston, H. *Environmental Ethics: Duties and Values in the Natural World*. Philadelphia: Temple University Press, 1988.

Rolston, H. "Wonderland Earth in the Anthropocene Epoch." In *Climate Change Ethics and the Non-Human World*, edited by B.G. Henning and Z. Walsh, 196–210. 2020.

Rose, G. *Visual Methodologies: An Introduction to Researching with Visual Materials*. London: Sage, 2016.

Russman, U. and J. Svensson. "Introduction to Visual Communication in the Age of Social Media: Conceptual, Theoretical and Methodological Challenges." *Media and Communication* 5, no. 4 (2017): 1–5. doi:10.17645/mac.v5i4.1263.

Ryder, R. "Speciesism Again: The Original Leaflet." *Critical Society* 2 (1970/2010). http://lists.exeter.ac.uk/items/8F806B86-A88B-950D-2658-579E23AC44AA.html

Scarborough, P., P.N. Appleby, A. Mizdrak, A.D.M. Briggs, R.C. Travis, K.E. Bradbury, and T.J. Key. "Dietary Greenhouse Gas Emissions of Meat-Eaters, Fish-Eaters, Vegetarians and Vegans in the UK." *Climatic Change* 125, no. 2 (2014): 179–92. doi:10.1007/s10584-014-1169-1.

Scudder, J.N. and C. Bishop-Mills. "The Credibility of Shock Advocacy: Animal Rights Attack Messages." *Public Relations Review* 35, no. 2 (2009): 162–64. doi:10.1016/j.pubrev.2008.09.007.

Singer, P. *Animal Liberation: A New Ethics for Our Treatment of Animals*. New York: Random House, 1990. (Original work published 1975.)

Steinfeld, H. *Livestock's Long Shadow: Environmental Issues and Options*. Rome: Food and Agriculture Organization of the United Nations, 2006.

Stenmark, M. "The Relevance of Environmental Ethical Theories for Policy Making." *Environmental Ethics* 24, no. 2 (2002): 135–48. doi:10.5840/enviroethics200224227.

Sullivan, N. and N. Longnecker. "Choosing Effective Frames to Communicate Animal Welfare Issues." In *11th International Conference on Public Communication of Science and Technology (PCST) in New Delhi, India—Archive*, 146–50. December 2010. https://pcst.co/archive/paper/811

Swim, J.K. and B. Bloodhart. "Portraying the Perils to Polar Bears: The Role of Empathic and Objective Perspective-Taking Toward Animals in Climate Change Communication." *Environmental Communication* 9, no. 4 (2015): 446–68. doi:10.1080/17524032.2014.987304.

Taylor, S. "Animal Crips." *Journal for Critical Animal Studies* 12, no. 2 (2014): 95–117.

Wisneski, D.C. and L.J. Skitka. "Moralization Through Moral Shock: Exploring Emotional Antecedents to Moral Conviction." *Personality and Social Psychology Bulletin* 43, no. 2 (2017): 139–50.

Worldwatch Institute. "Is Meat Sustainable?" *World Watch Magazine* 17, no. 4 (July/August 2004). www.worldwatch.org/node/549.

4 Suffering, sentientism, and sustainability

An analysis of a non-anthropocentric moral framework for climate ethics

Rebekah Humphreys

Introduction

Whilst many of the risks climate change poses to other than human life cannot be foreseen, some can be predicted, with probabilistic outcomes given to different climatic world scenarios in relation to a range of mitigation responses (see, for example, the most recent National Climate Assessment report released by the US Global Change Research Program on the current and future impacts of climate change, USGCRP 2017). Indeed, leading-edge research presented in *Climatic Change* (Warren et al. 2018) analyses the risks to "priority places" in the light of different possible futures, from an unmitigated case in which there are no cuts to emissions levels, to a mitigated case in which emissions are restricted to no more than 2 degrees centigrade above pre-industrial levels; the research indicates that up to half of the plant and animal species in these areas of significant diversity could face extinction by 2100 in the former case, with a 25 percent species loss even in the latter case. Thus even if a scenario in which the levels proposed by the Paris Agreement were met, there would be significant loss in biodiversity (Warren et al. 2018; WWF 2018), and already, we can see the impacts of climate change on some species (see Xu et al. 2009; Colwell et al. 2008; Parmesan 2006; Root et al. 2003).

However, in terms of the available literature, while there is some focus on species depletion and loss of biodiversity in relation to the impacts of climate change on other than human beings, little focus is given to *individual* animals and *their* flourishing or well-being. And as Palmer rightly claims (2016, 132), there is much more by way of considerations of the impacts on human beings. But in the case of such latter considerations, the discussion by comparison is very different—here, the fulfilment or thwarting of the interests of human beings in relation to climate change appear to be considered over and above the different (but related) considerations linked to the survival of human beings *as a species*. In the light of the largely anthropocentric outlook of the Western world, this is in some way to be expected; humans are deemed to have a higher moral status than non-human beings, often in virtue of what are thought to be uniquely human capacities (including capacities deemed to qualify most humans as persons and as individuals in their own right). In respect of climate change concerns, non-human creatures tend to be perceived as "belonging to a species" rather than as beings with individual interests, whether because of anthropocentric reasons relating to, for example, conservation, education, and the preservation of ecosystems essential for human survival or for ecocentrist reasons pertaining to appeals to the intrinsic good of ecosystems and species themselves.

Indeed, individual animals and their interests appear then to get "lumped" into the category to which they belong in terms of species membership, whilst humans appear to retain an elevated status in so far as they are perceived as having interests that matter apart from their contribution (or lack of) to the thriving or furthering of *Homo sapiens* – interests which are due moral consideration whether or not the continuance of the species is considered as a separate concern.

This dualism of our inherited philosophical legacy can be seen more prominently in our ideas regarding animals used in commercial practices, particularly intensive rearing, a practice in which billions of animals are made to endure pains, discomforts, and frustrations, as well as forced to live a life in which they are prevented from fulfilling their natural tendencies. It is beyond the scope of this chapter to discuss the attempted justifications for such treatment; it is sufficient to say here that such attempts usually appeal to farmed-animals' supposed lack of certain capacities (for discussion of the ethics of factory farming, see Humphreys 2014, 2010).

Whether or not we agree with the claim that such mistreatment of other than humans is partly related to our philosophical legacy, all sentient animals have interests – interests that are all too often not sufficiently considered by human beings. And with respect to an ethics of climate change mitigation in relation to sentient creatures, even if we do not meet the goal of the Paris Agreement, let alone curtail greenhouse gases at a more sustainable level, at least some non-human sentient creatures will survive and will probably outlive humanity, but if the interests of sentient beings matter in their own right, then this confers responsibilities to mitigate the devastating impacts of climate change for the sake of actual and future sentient creatures generally, not just for the sake of human beings only.

Sentientism

Sentientist philosophers have presented convincing arguments to show why the interests of sentient animals are of direct moral concern, and no more so than Peter Singer, who may be seen to exemplify the sentientist stance, at least with regard to his position on that which has moral standing and why. Indeed, while it is to be expected that the complexities of the views of each of such philosophers will diverge considerably (particularly in relation to the ethical theories they support regarding what makes right actions right, what they consider to be of intrinsic value and why, where they locate value or values, and how they deal with conflicts of interests), by definition, all such philosophers support the proposition that sentience is both a necessary and sufficient condition for having moral standing – that is, for having interests that are relevant in the moral arena. (For the purposes of

this chapter, it is sufficient to focus on this proposition with regard to a consideration of convergences among sentientist theorists, but it should perhaps be noted
that this is not the only proposition that would be supported by all adherents of a
sentientist viewpoint. See Rodogno (2010) for further discussion of other claims
shared by all sentientists.)

It is on this basis that Singer effectively presents his argument against that
which Richard Ryder first coined "speciesism"; unfair discrimination or an unfair
weighting of interests based on a morally irrelevant characteristic—in the case of
non-human beings, lack of membership of the species *Homo sapiens*, or lack of
supposedly uniquely human characteristics deemed necessary for inclusion in the
moral sphere (Ryder 1975, 1–14; Singer 1995, 1–23). For Singer, such speciesism
amounts to a failure to recognize sentient creatures as beings to which the principle of equal consideration of interests applies—indeed, in so far as animals have
the capacity to suffer, they do have interests, at the very least an interest in not
suffering (Singer 1995, 8). As such, the principle of equality, as a principle concerning our treatment of creatures with *interests*, applies not just to human beings
but non-human ones too and requires that *like* interests be considered equally (but
not that all sentient creatures be treated the same). This has serious implications
for an ethic of climate change (which will be discussed in the following).

Despite the soundness of Singer's argument for extending the principle of
equality to animals, there has been a move in animal ethics to attempt to give
sentient animals serious (but not necessarily equal) consideration via an analysis of the concept of personhood and how it might be applicable to animals – a
concept often bound to certain cognate capacities that are thought to allow for
a degree of self-awareness over time and planning for the future. Accordingly,
while sentience is necessary and sufficient for moral standing, it is not sufficient
to qualify as being as deserving of equal consideration. Gary Varner, for example, argues that the lives of persons, in their ability to develop a narrative story
of their lives, have interests that are more morally significant than the lives of
sentient non-persons (2011, ch. 6). And Singer himself considers there to be a
distinction between sentient persons and sentient non-persons (a distinction which
he believes, like Varner, makes the lives of the latter replaceable in a way that
the lives of the former are not), suggesting that while some animals are persons,
some humans are not persons (Singer 1995, 20–21). However, even if one denies
that some animals have personhood, some possess capacities that may be said
to enhance their phenomenological experiences (including, for example, olfac-
tory powers, echolocation, and more finely tuned perceptual faculties generally),
capacities that persons may possess to a much lesser degree. Such capacities may
add moral significance to the lives of some animals but not necessarily to persons
who lack such capacities.

Admittedly, there are good reasons for claiming that some lives are more valuable than others when interests conflict because some animals may be harmed
more by death than others due to, say, their greater or more complex capacities. And, after all, the principle of equality, applied correctly, does not require
equal treatment but equal consideration of *comparable* interests. But this is very

different from claiming that the lives of many sentient beings are replaceable for the very reason that their lack of certain capacities means that they are deemed non-persons. One problem with this is that if such beings have worthwhile lives, then to kill them is to injure them (Clark 1977, 59; see also Humphreys 2014). But besides this, the replaceability stance is susceptible to the argument from marginal cases if it is followed by an attempt to claim that while most sentient animals are non-persons and thus replaceable, all humans, even marginal ones, should be considered *as* persons (plausibly, this is a move that Singer does not make) – that is, should be considered as having lives that are more morally significant than sentient non-humans, even though some marginal humans may have little quality of life and few developed cognitive capacities compared with those animals that (although sentient) are not classed as persons. (For a discussion of the argument from marginal cases, see Singer 1995.)

Varner, for example, provides indirect reasons for giving all human lives equal moral significance whilst rejecting the claim that (in judgments of moral weight) such moral significance should be given to those animals with at least similar cognate capacities to marginal humans; these reasons pertain to the close relationships we form with humans and the fear that would be created if we treated marginal humans in the way we now treat other than human non-persons (Varner 2011, 253–54).

But this provides no direct ground or justification at a theoretical level for supposing that marginal humans should be given more moral significance than non-humans that are not classed as persons (for a more detailed response to Varner, see Attfield and Humphreys 2013). The principle of equal consideration of interests, applied to animals, requires that like interests be given equal consideration, and having an interest comparable to another being's interest is not necessarily dependent on species membership. In any case, we should be wary of slippage from the claim that the lives of certain beings are more valuable by virtue of their possession of certain capacities to the claim that when their interests conflict with beings who lack such capacities, the latter have less weighty interests than the former (in particular, wary of an erroneous move to the claim that supposed nonpersons have a less significant interest in not suffering than persons or that the interests of nonpersons in not suffering become less weighty when the other interests at stake are harnessed to beings that may be said to possess personhood).

That said, whether we grant personhood to certain animals, it is clear that with regard to climate change, the interests of all sentient other-than-human creatures stand to be affected. If their interests are to be given equal consideration to the like interests of humans, then this creates obligations to do much more than aim to meet the target of the Paris Agreement.

Some implications: climate ethics

In this way, sentientism could be seen to offer the environmentalist justifications for mitigations that would pay heed to the significant interests of sentient non-humans and give those interests equal consideration as the like interests of humans in climate change negotiations, with some sentientists claiming more generally that their normative stance can serve as an adequate environmental ethic (Jamieson 1998; Varner 2001).

One implication of this is that negotiators would have to ensure that the habitats of wild sentient creatures are sufficiently protected at least to the extent that such creatures can satisfy their basic needs (for a discussion of 'wildness' and its different meanings in the context of climate change, see Palmer 2016); such an implication would of course involve refraining from acting in ways that give less weight to like interests, even when our own interests are at stake and especially when those interests are peripheral. The fulfilment of the interests of wild animals depends on at least minimal interference (for further discussion, see Palmer 2015), and this would involve acting in ways that respect their habitats. Of course, we appear to be failing to avert the negative outcomes of climate change with respect to the homes of a staggering number of members of our own species, let alone individuals of other sentient species (it has been estimated that there could be as many as 1.4 billion environmental refugees by 2060 [see Environmental Justice Foundation 2016, 14]). But with regard to non-human creatures, the situation appears to be at least as dire, considering that the extent of climate refugia is already declining at an alarming rate and is set to decline even further (Warren et al. 2018). Indeed, as said earlier (in this section and at the beginning of the chapter), given that climate change is devastating wild populations (ibid.), then this creates positive obligations in respect of devising mitigation and adaption strategies that take the interests of wild creatures into account. (See further John Nolt's "Climate Change and the Loss of Non-human Welfare" (2019) for a discussion of harms in relation to wild animals in the context of climate change.)

And yet, in contrast to the aforementioned implication, some hold that it follows from sentientism that human beings should intervene to prevent predation (see Sagoff 1984; see further Sapontzis 1984), meaning that moral agents in general should intervene to prevent lions, eagles, and crocodiles catching and eating their prey. This would of course be contrary to any feasible attempt to combat environmental pressures resulting from climate change. But there is a persuasive case against such interventions in any case. For the well-being of prey species and of predators depends on the process of predation continuing, as does the very existence of predators; if there is value in the lives and the flourishing of predators as well as prey, then it should be allowed to continue. With regard to climate ethics, the continuing operation of natural systems should itself be permitted and necessitates predation.

Underlying the idea that sentientism or, more specifically, giving equal consideration to the interests of animals implies we should interfere to prevent predation is the assumption that the sentientist stance requires that we should prevent all animal suffering. But neither sentientism nor equality for animals requires this, no more than equality for humans requires this; to suppose that it does is to misunderstand what counts as ethical or unethical treatment. That animals suffer in the wild is part and parcel of the life and death cycles of wild animals, cycles on which the healthy functioning of ecological processes depend and which sustain all life. Besides, interference with predation would contribute to the total suffering in the world, rather than alleviate it (issues arising from the total view set aside, see Attfield 1995, ch. 10, for a full discussion of the total view), not least because it would prevent animals from fulfilling their natural tendencies and potentialities and result in some animals starving to death. This, in turn, would, of course, to a certain extent disrupt the ecological balance, which could have disastrous consequences for humans and sentient animals but would certainly undermine any sentientist attempts to tackle the impacts on animals in relation to climate change. Therefore, a coherent sentientist ethic of climate change either must reject the claim that all suffering is intrinsically bad or reject the claim that all suffering should be prevented or reject both.

With regard to sentient creatures, the welfare of whom is dependent on human beings, then there is a direct obligation to abolish such methods in any case. This particularly applies in the case of the billions of animals made to ensure severe and prolonged suffering through factory farming methods. We would never consider it justifiable to inflict comparable suffering on humans for the same purposes – for the mass production of meat, that is. (See preceding for a discussion of impacts of climate change on wild animals.) Accordingly, on the principle of equal consideration of interests, our treatment of factory-farmed animals is not (and will never be) permissible. In relation to climate change, the International Panel for Sustainable Resource Management reports that "Agriculture and food consumption . . . are one of the most important drivers of environmental pressures, especially habitat change, climate change, water use and toxic emissions" (UNEP 2010, 33), and "in the case of intensive agricultural process, growing can also be very polluting due to the use of fertilizers and pesticides. Agriculture also puts pressure on land and water use, as well as energy use" (UNEP 2010: 29), disproportionately effecting geographic areas that may be classed as "food-insecure" (see FAO 2017a, 2017b).

It would appear then that taking proper account of the interests of factory-farmed animals, in so far as it would involve at least a move towards abolition, would significantly reduce emissions for the good of all and could well release land for crops for human consumption, land which is currently used to grow soya crops to feed the farm animals reared for consumption, an incredibly wasteful practice (this would have further implications with regard to sustainable land use in relation to issues concerning poverty, malnutrition, and equitable global development, distribution problems notwithstanding). (For a further discussion of the ethics of factory farming, see Humphreys 2010, 2014; for research on the costs borne by humans in relation to factory farming, see Tansey and D'Silva 1999; Henning 2011).

Overall, then, the negative environmental impacts of intensive food systems in relation to climate change provide reasons to support the abolition of intensive methods of farming (reasons aside from appeals to the suffering endured by intensively reared farm animals). Further, these reasons need not be indirect in the context of sentientism, for such impacts *directly* affect sentient wild animals and human beings. Even though intensive rearing methods would be argued by sentientists to be unjustifiable in any case based on the suffering they cause (that is, irrespective of the environmental devastation it causes), human and non-human suffering included, such an argument in relation to climate change would provide another (albeit this time indirect) reason for moving towards more sustainable agricultural methods. Moreover, in the light of the principle of equal consideration—or even just the need for serious consideration—and given at the very least what we already know in relation to climate change and its harmful impacts on wild animals, there is a strong obligation to adopt targets for action that properly consider their species-specific interests.

Theoretical challenges

It seems, then, that sentientism (in relation to pressing environmental concerns), could not only afford genuine protection for and equitable treatment of sentient beings but also propose at least some actions that cohere with obligations to minimize the negative impacts of climate change on humans, non-humans, and ecosystems. And if the principle of equality is applied to animals, which plausibly it should be (Singer 1995), then such obligations are a matter of equity not just in relation to humans but non-humans too (Attfield and Humphreys 2016, 2017).

Of course a being's interest in not suffering is certainly a morally significant one in that its fulfilment or frustration can seriously affect its life for better or for worse respectively, seriously impacting on its quality of life (see Goodpaster for the distinction between moral standing and moral significance, 1978). Such an interest can often override other peripheral and even weighty interests (including, for example, an interest in continued existence, for there are some forms of existence that constitute a greater harm than death for the very reason that they cause too much suffering). Nevertheless, it is also reasonable to claim that we can harm not only sentient creatures but nonsentient ones too—which strongly indicates that the latter have interests. And such harms in some cases could be greater than the harms caused in relation to other conflicting interests at stake. In other words, the interests of nonsentient creatures could well outweigh the interests of sentient ones, depending on the case in question. Of course, sentientists would reject this proposition on the basis that nonsentient creatures do not have interests of a morally relevant kind.

But this poses a serious challenge for sentientists. Besides issues arising from where to draw the line for sentience (see, for example, Braithwaite 2010) and although there are many disanalogies between sentient and nonsentient life, there is an analogous argument (from the sentient case to the nonsentient one) for holding that nonsentient creatures have interests (even though they do not take an interest in their life) – at least interests in, for example, flourishing or thriving after their own kind (see Taylor 1986: 63–68; Attfield 2014, 12, 24–25, 49). And our responses to hypothetical scenarios derived from Last Man cases (Routley 1973), in particular the modified thought experiment presented by Attfield (1994, 168) whereby the Last Man needlessly chops down the last tree (a tree that could multiply if left alone), fail to cohere with the sentientist's claim that there is nothing of value at stake here, as do responses to other thought experiments such as that of Planet Lifeless and Plant Flora, put forward by Donald Scherer (1983). (See further Attfield's "Biocentrism, Climate Change, and the Spatial and Temporal Scope of Ethics" [2020] for an enlightened discussion of these thought experiments.) With regard to an ethic of climate change, we could easily imagine large tracts of biodiverse areas that do not contain sentient life but do contain many other living things. But sentientists provide no direct reason for protecting such habitats if doing so does not further the interests of sentient creatures. Moreover, if we think that what the last man destroys is of value or that in destroying Planet Flora we do wrong or that in destroying large tracts of land containing much nonsentient life, we do harm, then we have already moved beyond the bounds of sentientism towards a more inclusive approach to environmental concerns that embraces nonsentient (as well as sentient) life. While the sentientist may well afford some protection for the habitats of nonsentient beings—she could, similarly to Jamieson (1998; see also Crisp's response to Jamieson 1998), appeal to the benefits of such habitats in relation to the well-being of all sentient creatures, following a Norton-like (Norton 1991) yet (in contrast) a nonanthropocentric argument—whether such protection would be sufficient is a different matter.

Overcoming limitations: ecocentric values

One move here is to consider whether only some form of ecocentrism would be an appropriate grounding for an environmental ethic and thus whether we should locate intrinsic value in ecosystems and collective entities and whether we should consider them as bearers of interests. But while moral standing belongs to all and only beings with interests, for all and only such beings are capable of being harmed and benefited or having their own good (see Goodpaster 1978; Feinberg 1974), it is far from obvious that such entities do have interests distinct from the individuals such systems support or the individuals that constitute such entities. Moreover, we should resist the suggestion that systems have moral standing and independent value, for the "good" of an ecosystem may be said to consist rather in the good of its members as individuals. (See further Nolt [2019] and Attfield [2019] for convincing reasons to suppose that it is individuals that have moral standing, as opposed to species.) Of course, those things that are of value do exist in the biosphere, and their existence does depend on the biosphere and its systems as a whole. But it does not follow that the biosphere and its systems are of intrinsic value.

Defenders of ecocentrism also support the view that (as well as systems) species have intrinsic value and moral standing. However, the term *species* denotes an abstract if not general notion identifying a category or set of individual organisms consisting of genetically like individuals, and though we can talk in a metaphysical sense of the "flourishing" of a species (and such talk may have rhetorical value), a species in and of itself does not have interests or intrinsic value. Rather it is the individual members of a species that have interests and the well-being or flourishing of individuals that has intrinsic value. Just as a crowd does not count as having moral standing because it is the individuals which make up the crowd that have that standing, the same is true of a species group. What is wrong about eliminating a species is that it cuts off future members of that species, and future members of a species do, or rather would, have a good of their own and moral standing (Attfield 1995, 24–25, 2014, 39). Further, just as we have obligations regarding future humans, whoever they may be (that is, without knowing their identity [see Parfit 1984, ch. 16]), we also have obligations towards future creatures, whatever their identity may be.

In relation to climate ethics, then, while the impacts of climate change may be said, for example, to result in species becoming extinct, it is the individuals (future and present) of those species that stand to be affected by any mitigation and adaptation strategies we adopt. That said, individuals may be susceptible to present and future harms related to climate change for the very reason that they belong to a particular group that is under threat. But, as Palmer plausibly claims in relation to one type of argument involving an analysis of group "harms," "it's not something *other* than the individuals that is harmed" (2009, 596).

That said, overcoming the limitations of a sentientist account of value need not involve a move towards holism. Indeed, while it may indeed be "a fact that environmentalists tend to think like holists" (Varner 2001, 201), we can take account of the good of all living beings, both sentient and nonsentient, without being required to endorse the view that classes of things (denoted by general terms, such as "species" or "systems") have interests.

Of course, much will depend here on one's conception of what counts as a morally relevant interest—for the holist direct moral relevance is tied to groups of kinds such as species and ecosystems. But as Varner argues,

> Holists must either show how species and ecosystems have interests in some traditional sense, or give a convincing account for attributing intrinsic value to them on some other basis. But since holists universally reject the first option, the burden of proof is on the holist to explain why such entities have intrinsic value.
>
> (2001, 202)

Besides, endorsing a form of holism would fail to account adequately for the individual interests of creatures currently affected or which stand to be affected by climate change. With regard to the sentientist's concept of a morally relevant interest, however, direct moral relevance is tied to sentience, but (as noted in the previous section) the possession of sentience is not a necessary condition for moral standing (although it may be sufficient). Arguably, while sentience is a characteristic that can be considered as morally significant in cases of conflicting interests, all nonsentient creatures have interests in thriving after their own kind, exercising their species-specific tendencies, and fulfilling potentialities.

Values for a sustainable world

And yet any policies regarding mitigation and adaptation will probably be human-centered ones, focused on reducing emissions for the sake of human beings only.

Of course, there is a strong case for saying that much more strenuous efforts need to be made in this regard in any case. But human beings may well be outlived on this planet by much nonsentient and possibly sentient life, and if we think that the Earth inhabited by such life has a value even in the absence of human beings, then this creates obligations to act in the interests of not just human beings but non-human ones too and not just sentient beings but also nonsentient ones (see Attfield and Humphreys 2016, 8). To illuminate the implied difference here between an anthropocentric approach and a nonanthropocentric one, the author draws attention to the rather significant difference between restricting emissions to no more than 2 degrees centigrade (as agreed in Paris) and reducing the amount of CO_2 in the atmosphere to no more than 350 parts per million (as agreed by climate scientists; see, for example, Hansen et al. 2008). The former has and would involve mass extinction. The latter is what would be required to return us closer to our Holocene "normal" and would be good for human and non-human life.

Admittedly, anthropocentrism may advocate similar environmental practices and policies to nonanthropocentric theories (Norton 1991, 237–43; see also Norton 2008) and attempts to tackle the mammoth problems facing humans in relation to climate change will, most probably, be tackled by a sensitive Norton-like theory. But this is not to say that those policies would be similar in terms of providing reasons for action—and it is these reasons which ultimately will provide the defense for our actions and which will themselves determine whether outcomes are sustainable not just for human beings but for other creatures too.

As Katie McShane argues, the central normative claim of anthropocentrism

> is not a claim about how we ought to behave. It is a claim about which features of non-human things can make them matter in which ways . . . Claims about why something has value are claims about why we, as moral agents, have a reason to care about the thing.
>
> (2007, 172; see also McShane 2008)

This is also true of nonanthropocentric claims; indeed, they are claims about what has value and why. Thus, if anthropocentrism and nonanthropocentrism do propose the same actions, then they will do so for very different reasons based on diverging claims as to what has independent value and thus moral standing: "claims about why we as moral agents should care about a thing serve as the grounds for ethical norms concerning the thing" (McShane 2007, 173). Since it is not the case that only humans have interests or that only human interests are important in terms of proposals regarding mitigation and adaptation strategies, then we should be wary of the implications and consequences of a theory that considers justifications for actions to be defensible if they appeal directly to human interests only. Forms of enlightened anthropocentrism may well propose actions that are right, but if the reasons given for those actions are discriminatory or unjustifiable, then it is not sustainable as a theory that aims, in practice, to tackle pressing environmental problems, problems for which our discriminatory or unjustifiable attitudes towards the non-human world are at least partly responsible in any case. In the light of this, we would be prudent not to foster such attitudes through application of a theory that values only human interests.

Similarly, we should be wary of promoting sentientist policies that are based on a direct consideration of the interests of sentient creatures only. Such policies may propose some similar practices to policies based on a direct consideration of *both* the interests of sentient and nonsentient creatures. But sentientist policies would fail to recognize sufficiently those interests the fulfilment (or even mere consideration) of which are not necessarily in the interests of sentient beings or not necessarily tied to their flourishing, including the interests of nonsentient creatures who may well outlive human beings (even if we manage to reduce emissions to a level feasible for the temporal continuance of much sentient human and non-human life). If there is value in a world of nonsentient life in the absence of sentient life, then whether we endorse sentientist or alternative nonanthropocentric policies instead becomes of utmost importance in terms of proposing not only defensible but sustainable climate change policies in the interests of all those creatures that are impacted by environmental pressures.

Conclusion

To conclude, although sentientism offers a less exclusive and more creditable ethic upon which to base an ethic of climate change than anthropocentrism, it is not clear that it can provide sufficient protection for nonsentient creatures and their environment. On sentientism, the nonsentient world is protected only so far as it promotes the interests of sentient beings. But there is much nonsentient life which has so far not been identified (including most of the life that inhabits biodiverse habitats). Just as this life cannot all be said to promote human interests (Attfield 2014, 76), it also cannot all be said to promote the interests of sentient beings generally, and, therefore, sentientism (similarly to anthropocentrism) provides no sufficient or direct grounds for protecting the undiscovered, nonconscious terrestrial and aquatic plant and animal life the existence of which is a known unknown. Not only this, but if most people do not care about certain habitats or nonsentient creatures or believe that the protection of certain nonsentient life will not promote their interests, then it is unlikely that sentientism can provide sufficient grounds for protecting those animals and habitats, for sentientist theories imply that we need to be restricted in our use of the environment only if such use does not adversely affect sentient individuals.

An adequate environmental ethic upon which to base climate change mitigation policies needs to recognize the moral standing of all individual creatures, sentient and nonsentient (for further discussion relating to the criterion for moral standing, see Routley 1973; Goodpaster 1978; Attfield 2014). It will be clear by now that the author supports an egalitarian biocentrist stance, not egalitarian in the sense of Taylor's position which claims that living things have equal inherent worth (Taylor 1986, 75), but in the sense in which like interests should be given equal consideration, whether those interests belong to sentient creatures or nonsentient ones. In relation to an ethic of climate change, biocentrism would endorse far more stringent mitigation policies than sentientism, not least because in recognizing the moral standing of all living beings, it would consider the interests of all beings to have direct moral relevance (for further discussion on biocentrism and climate change, see Attfield 2017; see further Humphreys 2016). But the author will not rehearse a discussion of this form of biocentrism and its implications, as this has been presented elsewhere (Humphreys 2016; see also Attfield and Humphreys 2016, 2017). Suffice it to say here that while sentientism would afford much greater protections for sentient creatures than anthropocentrism if it were used as a foundation for a climate change ethic (for a discussion of the inclusion of the interests of at least some non-human beings in climate change policy, see McShane 2016), it cannot account for the challenges posed by considering that even in scenarios in which stricter mitigation policies were adopted than those proposed by the Paris Agreement, the interests of most of the world's living creatures (nonsentient beings), present and future ones included, would fail to be included as interests that should be given direct consideration with regards to appropriate policies and practices. This is unjustifiable if the nonsentient world has a value apart from the interests of human and non-human sentient beings, which surely it does. Such a conclusion in respect of climate ethics calls for a proposed target for action that would be good for all living things and thus a much more ambitious target (of a reduction of CO_2) to no more than 350 parts per million, rather than that agreed to in Paris.

References

Attfield, Robin. "Biocentrism, Climate Change and the Spatial and Temporal Scope of Ethics." In *Climate Change Ethics and the Non-Human World*, Routledge Research in the Anthropocene, edited by Brian Henning and Zack Walsh. London: Routledge, 2019.

Attfield, Robin. "Climate Change, Environmental Ethics, and Biocentrism." In *Climate Change and Environmental Ethics*, edited by Ved P. Nanda, 31–41, 2nd ed. New York and London: Routledge, 2017.

Attfield, Robin. *Environmental Ethics: An Overview for the Twenty-First Century*. 2nd ed. Fully revised and expanded. Malden, MA and Cambridge, UK: Polity Press, 2014.

Attfield, Robin. "The Good of Trees." In *Environmental Philosophy: Principles and Prospects*. Aldershot: Avebury, 1994.

Attfield, Robin. *Value, Obligation, and Meta-Ethics*, Value Inquiry Book Series, vol. 30. Amsterdam and Atlanta, GA: Rodopi, 1995.

Attfield, Robin and Rebekah Humphreys. "Justice and Non-Human Beings, Part I." *Bangladesh Journal of Bioethics* 7, no. 3 (2016): 1–11.

Attfield, Robin and Rebekah Humphreys. "Justice and Non-Human Beings, Part II." *Bangladesh Journal of Bioethics* 8, no. 1 (2017): 44–77.

Attfield, Robin and Rebekah Humphreys. "Review of Gary E. Varner. In *Personhood, Ethics and Animal Cognition: Situating Animals in Hare's Two-Level Utilitarianism*. Oxford and New York: Oxford University Press, 2012." *Philosophy* 88, no. 345 (2013): 493–98.

Braithwaite, Victoria. *Do Fish Feel Pain?* Oxford: Oxford University Press, 2010.

Clark, Stephen R. L. *The Moral Status of Animals*. Oxford: Clarendon Press, 1977.

Colwell, Robert K., Gunnar Brehm, Catherine L. Cardelús, Alex C. Gilman, and John T. Longino. "Global Warming, Elevational Range Shifts, and Lowland Biotic Attrition in the Wet Tropics." *Science* 322, no. 5899 (2008): 258–61.

Crisp, Roger. "Animal Liberation Is Not an Environmental Ethic: A Response to Dale Jamieson." *Environmental Values* 7, no. 4 (1998): 476–78.

Environmental Justice Foundation (EJF). *Beyond Borders: Our ChangingClimate—Its Role in Conflict and Displacement*. A Report Produced by the Environmental Justice Foundation. London: EJF, 2016.

Feinberg, Joel. "The Rights of Animals and Unborn Generations." In *Philosophy and Environmental Crisis*, edited by W. T. Blackstone, 43–68. Athens, GA: University of Georgia, 1974.

Food and Agriculture Organization of the United Nations (FAO). *The Future of Food and Agriculture: Trends and Challenges*. Rome: FAO, 2017. www.fao.org/publications.

Food and Agriculture Organization of the United Nations (FAO). *The State of Food and Agriculture: Leveraging Food Systems for Inclusive Rural Transportation*. Rome: FAO, 2017. www.fao.org/publications.

Goodpaster, Kenneth. "On Being Morally Considerable." *Journal of Philosophy* 75, no. 6 (1978): 308–25.

Hansen, J., Mki Sato, P. Kharecha, D. Beerling, R. Berner, V. Masson-Delmotte, M. Pagani, M. Raymo, D. L. Royer, and J. C. Zachos. "Target Atmospheric CO2: Where Should Humanity Aim?" *Open Atmospheric Science Journal* 2 (2008): 217–31.

Henning, Brian. "Standing in Livestock's 'Long Shadow': The Ethics of Meat Eating on a Small Planet." *Ethics and the Environment* 16, no. 2 (2011): 63–93.

Humphreys, Rebekah. "The Argument from Existence, Blood-Sports, and 'Sport-Slaves'." *Journal of Agricultural and Environmental Ethics* 27, no. 2 (2014): 331–45.

Humphreys, Rebekah. "Biocentrism." In *Encyclopedia of Global Bioethics*, 263–72. Dordrecht, The Netherlands: Springer, 2016.

Humphreys, Rebekah. "Game Birds: The Ethics of Shooting Birds for Sport." *Sport, Ethics and Philosophy: Journal of the British Philosophy of Sport Association* 4, no. 1 (2010): 52–65.

Jamieson, Dale. "Animal Liberation Is an Environmental Ethics." *Environmental Values* 7 (1998): 41–57.

McShane, Katie. "Anthropocentrism in Climate Ethics and Policy." *Midwest Studies* 40, no. 1 (2016): 189–204.

McShane, Katie. "Anthropocentrism vs Nonanthropocentrism: Why Should We Care?" *Environmental Values* 16, no. 1 (2007): 169–85.

McShane, Katie. "Convergence, Noninstrumental Value and the Semantics of 'Love': Reply to Norton." *Environmental Values* 17, no. 1 (2008): 15–22.

Nolt, John. "Climate Change and the Loss of Non-Human Welfare." In *Climate Change Ethics and the Non-Human World*, Routledge Research in the Anthropocene, edited by Brian Henning and Zack Walsh. London: Routledge, 2020.

Norton, Bryan. "Convergence, Noninstrumental Value and the Semantics of 'Love': Comment on McShane." *Environmental Values* 17, no. 1 (2008): 5–14.

Norton, Bryan. *Towards Unity Among Environmentalists*. Oxford: Oxford University Press, 1991.

Palmer, Clare. "Against the View That We Are Normally Required to Assist Wild Animals." *Relations* 3, no. 2 (2015): 203–10.

Palmer, Clare. "Climate Change, Ethics, and the Wildness of Wild Animals." In *Animal Ethics in the Age of Humans*, The International Library of Environmental, Agricultural and Food Ethics, vol. 23, edited by B. Bovenkerk and J. Keulartz, 131–50. Dordrecht, The Netherlands: Springer, 2016.

Palmer, Clare. "Harm to Species—Species, Ethics, and Climate Change: The Case of the Polar Bear." *Notre Dame Journal of Law, Ethics and Public Policy*, Symposium on the Environment 23, no. 2 (2009): 587–603.

Parfit, Derek. *Reasons and Persons*. Oxford: Clarendon Press, 1984.

Parmesan, C. "Ecological and Evolutionary Responses to Recent Climate Change." *Annual Review of Ecology, Evolution, and Systematics* 37 (2006): 637–69.

Rodogno, Raffaele. "Sentientism, Wellbeing, and Environmentalism." *Journal of Applied Philosophy* 27, no. 1 (2010): 84–99.

Root, Terry L., Jeff T. Price, Kimberly R. Hall, Stephen H. Schneider, Cynthia Rosenzweig, and J. Alan Pounds. "Fingerprints of Global Warming on Wild Animals and Plants." *Nature* 421 (2003): 75–60.

Routley, Richard. "Is There a Need for a New, an Environmental, Ethic?" In *Proceedings XV World Congress of Philosophy*, 205–10. Bulgaria: Varna, 1973.

Ryder, Richard D. *Victims of Science: The Use of Animals in Research*. London: National Anti-Vivisection Society, 1975.

Sagoff, Mark. "Animal Liberation and Environmental Ethics: Bad Marriage, Quick Divorce." *Osgood Hall Law Review* 22, no. 2 (1984): 297–307.

Sapontzis, Steve. "Predation." *Ethics and Animals* 5, no. 2 (1984): 27–38.

Scherer, Donald. "Anthropocentrism, Atomism and Environmental Ethics." In *Ethics and the Environment*, edited by D. Scherer and T. Attig, 73–81. Englewood Cliffs, NJ: Prentice-Hall, 1983.

Singer, Peter. *Animal Liberation*. 2nd ed. New York: New York Review/Random House, 1995.

Tansey, Geoff and Joyce D'Silva, eds. *The Meat Business: Devouring a Hungry Planet*. London: Earthscan Publications, 1999.

Taylor, Paul. *Respect for Nature: A Theory of Environmental Ethics*. Princeton, NJ: Princeton University Press, 1986.

United Nations Environmental Panel (UNEP). *Assessing the Environmental Impacts of Consumption and Production: Priority Products and Materials*. A Report of the Working Group of the Environmental Impacts of Products and Materials to the International Panel for Resource Management. Lead Author: Edgar Hertwich, 2010.

U.S. Global Change Research Program (USGCRP). *Climate Science Special Report: Fourth National Climate Assessment*. Edited by D. J. Wuebbles, D. W. Fahey, K. A. Hibbard, D. J. Dokken, B. C. Stewart, and T. K. Maycock. Vol. I. Washington, DC: USGCRP, 2017.

Varner, Gary. *Personhood, Ethics and Animal Cognition: Situation Animals in Hare's Two-Level Utilitarianism*. New York: Oxford University Press, 2011.

Varner, Gary. "Sentientism." In *A Companion to Environmental Philosophy*, edited by D. Jamieson, 192–203. Malden, MA and Oxford, UK: Blackwell, 2001.

Warren, R., J. Price, J. Van Der Wal, S. Cornelius, and H. Sohl. "The Implications of the United Nations Paris Agreement on Climate Change for Globally Significant Biodiversity Areas." *Climatic Change* 147 (2018): 395–409.

World Wildlife Fund (WWF). "Half of Plant and Animal Species at Risk from Climate Change in World's Most Important Natural Places." March 14, 2018. Accessed July 15, 2018. www.worldwildlife.org/press-releases/half-of-plant-and-animal-species-at-risk-from-climate-change-in-world-s-most-important-natural-places.

Xu, Jianchu, Edward Grumbine, Arun Shrestha, Mats Eriksson, Xuefei Yang, Yun Wang, and Andreas Wilkes. "The Melting Himalayas: Cascading Effects of Climate Change on Water, Biodiversity, and Livelihoods." *Conservation Biology* 23, no. 3 (2009): 520–30.

5 Biocentrism, climate change, and the spatial and temporal scope of ethics

Robin Attfield

Introduction

In the course of arguing that we should take seriously the interests of generations living after our own deaths, Cicero once wrote as follows:

> As we feel it wicked and inhuman for men to declare that they care not if when they themselves are dead the universal conflagration ensues, it is undoubtedly true that we are bound to study the interest of posterity also for its own sake.[1]

As at the time of Cicero, it is widely accepted in our own day that the kind of egoistic approach which declares "Après moi, le déluge" (a remark attributed to the French King Louis XV) is unacceptably blinkered and myopic, and also that the good of future generations matters in its own right.

Such egoism, however, appears less unacceptable when it adopts a collective form and declares that what happens after the demise of humanity is no concern to us. After all, adherents of this collective variety of egoism can console each other with the thought that few if any of their concerns and commitments will be honored in that humanly unpopulated world. This is a kind of communitarian consolation, which pretends that none but those who share or will share our concerns and commitments matter. But this stance is rightly rejected by those who hold the more universalist stance of the playwright Terence, who had one of his characters say "homo sum, et nil humani alienum a me puto" (I am human and consider nothing human alien from myself). This includes, of course, human beings who have different concerns and commitments from our own.

Yet collective egoism, even when it embraces the whole of humanity in the manner of Terence, remains ethically blinkered, or so I shall be arguing. It may seem unfair to represent it as a form of egoism, as its adherents can cherish human company, society, and culture and need not lack respect for other people. The label of "anthropocentrism" appears to make this stance seem more humane. But if it forecloses concern for creatures of the post-human future and for promoting their good and alleviating their suffering through action taken in the present, then it is morally close to present neglect of contemporary animals and not far removed from cruelty towards them. But such neglect and cruelty, widespread as they may be, are almost universally condemned, and to a similar extent as that of the condemnation of unheeding egoism.

At one time, neglect of creatures of the post-human future was acceptable because there was little or nothing that human agents could do that would affect such creatures. A possible exception was the extinction of species, but in the days of Terence or of Cicero, there was no ability to monitor whether species had become extinct, as opposed to their no longer being seen in one's locality. But as Hans Jonas has said, in our own day, the scope of ethics has expanded, because human technology has made possible impacts of our actions (and inaction) that spread much further across space and time than ever before.[2]

To paraphrase Jonas, when the classical texts of ethics from Plato to Kant were written, the impacts of human action were seen as affecting almost exclusively the human contemporaries of the agent, and any long-term outcomes could be disregarded as serendipitous and unpredictable side effects, inessential for purposes of constructing adequate theories of virtue or duty. But now, because of technology, the impacts of a great deal of human action have to be recognized as affecting large swathes of the biosphere and future generations for many centuries to come.[3] Besides (we might add), anthropogenic climate change is disrupting many ecosystems and driving numerous species to migrate polewards, such that often they run out of habitats as they do so. While Jonas may not have been aware of influences such as this one, he was well aware of human impacts on wetlands, watersheds, rivers, and oceans.

This recognition led Jonas to reject any form of anthropocentric ethic. Jonas was aware of the many ways in which technology and the spread of human settlements are affecting non-human species and their prospects of survival into the future and felt that the scope of ethical concern should expand to match this increased sphere of influence and impact. While he was greatly concerned about the human future, his concern was not limited to that. Certainly, if non-human creatures are regarded as having moral standing, as was argued by Kenneth Goodpaster in 1978,[4] then the case for this expansion of ethical concern is compelling.

The spatial and temporal scope of ethics

How far in space and in time does the enlarged sphere of human influence extend? In the first place, it extends to the entire surface of our planet, including the depths of the oceans.[5] This now includes the strange creatures which flourish in the sulfurous vents that well up from the ocean floor, and it probably soon will include the creatures that live beneath the Antarctic ice shield. More generally, it extends to all those species (probably a majority) that have not yet been discovered or identified but whose survival is frequently under threat, such that they are liable to disappear before even being discovered.

Indeed, the spatial extent of human influence reaches at least as far as the full extent of the solar system, including its planets, moons, and asteroids. We have already littered this zone of space with the unwanted by-products of space-probes; the main ethical relevance of this spatial pollution lies in the possibilities of collisions in space and of harm to terrestrial life if any of this debris returns to Earth. But if our focus is on the living creatures liable to be affected by human influence, then these could include any creatures possibly living on Europa, a moon of Jupiter and, with somewhat greater probability, the creatures possibly inhabiting Enceladus, one of the moons of Saturn. And if living creatures have moral considerability, as Goodpaster argued, then the scope of our moral concern must extend to all these habitats or potential habitats, because of their living inhabitants.

I turn now to the temporal extent of human influence. At one time, it could safely be assumed that our successors would inherit much the same environment as ourselves, but this is no longer the case, granted the scale and speed of ecological change. We can now foresee the occurrence of extreme weather events of increasing frequency and intensity across the coming centuries, driven by climate change and ranging from floods and storms to droughts and wildfires, with widespread impacts both on human and non-human life and well-being.

Yet the effects of anthropogenic extinctions will last longer than this, as the extinct species are likely to remain extinct (unless reconstituted from fragments of DNA) across the whole of future human history, and after the eventual demise of humanity too. Indeed, some species could well outlive humanity unless we extinguish them in the coming decades. Thus, the influence of current human action and inaction can already be foreseen to extend for millions of years into the future, for the future of humanity could persist for as long as that and be deprived in all those generations of species extinguished in the present, and if humanity eventually becomes extinct, the posthuman biota of the planet will either include or exclude whatever species could be extinguished in coming decades or could be spared and live on beyond the human epoch into posthuman times.

Further, if it is not just the intended impacts of human action for which we are responsible, but also the foreseeable impacts (as I have argued elsewhere),[6] then our moral responsibilities cover the same period of time, as long as creatures with moral standing are affected for better or for worse. Clearly, given Goodpaster's account of "moral considerability" (a phrase synonymous in his terminology for moral standing), then this condition is satisfied, as those creatures that could have outlived humanity will be among those with moral standing. If so, then the temporal extent of our moral responsibility does actually extend to the period beyond the furthest reaches of human history.

Anthropocentric defenses

What can anthropocentrism say in the face of these apparently significant problems about its credibility? One defense is that a reason for preserving species and habitats consists in the value of scientific research and its centrality for human interests. However, there will be no scientific research after the demise of humanity, and so this defense is inapplicable to the case for preserving species such that they survive humanity and continue to flourish in our absence. Besides, scientific curiosity is selective, and preserving (say) areas of tropical forests and of coral reefs because they are rich in species, many of which have yet to be identified, could well fail to preserve less species-rich regions, such as boreal forests and tundra. If non-human species were preserved for this reason alone, the map of preserved areas would be patchy, with many areas rating too low a priority to qualify for science-based preservation.

Another argument might be the need to preserve all those ecosystems on which human beings depend. This argument admittedly upholds the preservation of many areas to which these anthropocentric reasons are relevant; the preservation of watersheds would be a leading example. But the value that this argument ascribes to non-human beings is instrumental, and this value would have to be weighed against the instrumental value of the same creatures deriving from their potential consumption by human beings for food, drink, or fuel or of disrupting them for the sake of the extraction of minerals. The upshot is once again a patchy world map, and one liable at that to be re-drawn each time a government or corporation decides to embark on some new project of exploitation. The resultant instrumental value of many non-human creatures would be either too slight to secure their survival or too negligible for anyone to care, if this were the only ground for caring.

Mark Sagoff has ingeniously suggested that the true basis for valuing non-human creatures is their symbolic value. Thus eagles and free-flowing rivers express American values and should be preserved for that reason.[7] One problem with this approach is that most creatures would be omitted from consideration through failing to symbolize whatever was locally valued. Another is that different values would be symbolized in different places, and so (for example) species that migrate (such as geese, storks, and swallows) might be treasured in one or more of their seasonal habitats but used as target practice in others, where their symbolic significance goes unrecognized. Yet another is the fluctuation over time of the values that even one and the same country celebrates. Hence the preservation of currently resonant and symbolic species might prove to be no more than temporary. (I have replied further to this basis for preservation elsewhere.)[8]

Anthropocentrists further need to reconcile their stance with the near-universal condemnation of cruelty towards animals and neglect of the animals in one's charge. There are two traditional defenses. One is that these forms of behavior are wrong because they damage human property. But this is no reply where the wrongness of harming one's own animals is concerned, whether through cruelty or through neglect. Nor is it a reply to the related widespread judgment that such cruelty would be wrong towards animals belonging to someone else even if no financial loss were to be suffered by the owner and even if, in cases where there is financial loss, compensation is paid.

The other defense is that cruelty (and so on) towards animals is prone to make its perpetrators more prone to be cruel (and so on) towards human beings. But this defense is based on a testable empirical claim, and experience does not invariably verify it. For while there must be some cases where it fits the facts, there are almost certainly others where people manage to compartmentalize their lives and after a day of cruel treatment of animals return to their families and contrive to behave as model spouses and parents.[9] Thus these defenses fail.

Some anthropocentrists, however, attempt to disarm criticism by claiming that value attaches not only to the objects of human interests but also to the objects of human caring (and simply because of this caring). Thus inflicting suffering on animals is wrong because many people care about it, just as the defacement of works of art is held to be wrong because it ruins the aesthetic enjoyment of those works on the part of people who would otherwise enjoy them, and thus can be said to care about their intactness.

This defense involves, of course, a modification of anthropocentrism, introducing offense to humans as a ground for attitudes like condemnation on a par with harm towards humans. Many Millian liberals have objected to such an extension, in the cause of promoting toleration of offensive behavior. But let us imagine or pretend that there is no problem about this modification and investigate whether it allows anthropocentrists to object to the same range of behaviors as most people adopt towards, for example, cruelty towards animals and the extinction of species.

Much depends here on how widespread the human caring has to be to make its focus objectionable. With universal caring, the implications would at least be clear. But what about cases where a majority care in one or two places, but there is no such majority elsewhere? Factory farming might comprise an example of a practice objected to in some countries but accepted in others. Are anthropocentrists to declare that this practice is unacceptable and wrong only where a majority object to it and acceptable everywhere else? This, however, clashes with the universal judgments that most people hold in such matters; if factory farming, it is usually held, is wrong in one country, then it is wrong in others as well. Anthropocentrists of the modified variety seem unable to uphold judgments of this kind, unless they adopt the more extreme position that it is wrong to behave in such a way as to conflict with *anyone's* concern and caring.

But if this more extreme modification is held, then new problems arise. For people's concerns frequently conflict, and the caring of some is often matched with caring to contrary effect on the part of others. For example, fox-hunting is a matter of concern for many in Britain, but opposition to fox-hunting is a matter of possibly equal concern for many others there. But, this being so, anthropocentrists appear to be committed to pairs of mutually conflicting judgments, for they appear to be committed both to fox-hunting being wrong and to objecting to fox-hunting being wrong, the latter indirectly implying that fox-hunting is acceptable and not wrong at all. And when a position can only be defended where its defense has such conflicting implications, the time has surely come for it to be abandoned.

It should be added that there are cases where species are extinguished without anyone caring, because the species in question has not even been discovered. While most people believe that causing such extinctions is at least highly undesirable and, unless there are countervailing factors, often actually wrong, these judgments cannot be allowed for even by modified anthropocentrism of the most extreme variety.[10] They require instead the adoption of a non-anthropocentrist ethics.

Astrobiology and planetary thought-experiments

As Donald Scherer has argued, certain thought-experiments suggest that widespread judgments transcend not only anthropocentrism but sentientism as well

— the normative theory, that is, that moral standing is confined to sentient organisms and moral significance to their well-being. To test this view, we need to imagine that there is a planet somewhere, whether in the solar system or beyond, which supports creatures capable of photosynthesis, of self-maintenance, of responding to their environment and of reproduction, but incapable of sensation or locomotion. Scherer calls this planet "Flora" and contrasts it with another planet that he calls "Lifeless," which sustains no such organisms (nor any others).

Scherer then asks whether we could morally accept a plan to destroy either of these planets, maybe for the sake of scientific knowledge or maybe with a view to retrieving the minerals released through shattering them. Perhaps we should add to his scenario that these planets are too distant to be seen either with the naked eye or through a telescope and that accordingly no human beings are going to suffer aesthetic loss through the destruction of either of these planets.

As Scherer proceeds to claim, most people would have little or no objection to the destruction of Lifeless. Here, we might add the proviso that its destruction was not going to ruin the cycles of the stellar system to which it belongs or any life forms that may be present there, in the way that the destruction of our moon would significantly harm life on Earth through the loss of the tidal system that depends on it. But most people, as Scherer remarks, would object to the destruction of Flora.[11]

This widespread and deep-seated judgment suggests that we regard the continuing existence of plants as valuable. Sentientists could imaginably argue that we do so only because there is a possibility for living organisms to evolve into sentient forms of life (despite this seemingly not having happened in the course of the evolution of life on Earth). But even if the thought-experiment is adjusted, in such a way to guarantee that this is not going to happen (maybe through there being a strong prospect of this planet being destroyed by a collision with an asteroid before such evolution could possibly take place), the widespread and deep-seated judgment remains unaffected.

Accordingly, we seem to regard the existence or perhaps the flourishing of plants as valuable just as they are, whatever their prospects for evolving into something else. And this judgment is all of a piece with Richard Routley's Last Man thought-experiment, where we imagine the sole survivor of a nuclear explosion that has destroyed all human life randomly destroying the trees around him. Most people judge this action to be wrong and continue to do so even if it is added that all other sentient creatures have been destroyed by the nuclear explosion and therefore cannot benefit from the Last Man's actions.[12] In response to this thought-experiment, some say that what we object to is the Last Man's vandalism. But this is an incomplete response, as there has to be something that makes vandalism to be bad, and that seems to consist in the destruction of the trees and of the good of their continued existence.

These thought-experiments count against both anthropocentrism and sentientism and suggest that most of us are already committed either to biocentrism, which counts all living creatures as having moral standing and their well-being as having independent value, or to ecocentrism, which regards species and ecosystems as having moral standing and/or independent value, either as well as all living creatures or instead of. Here, it is appropriate to add that both Hans Jonas and Kenneth Goodpaster appear to have been moving towards one or other of these stances. It will be clear from this section that the same applies both to Richard Routley and to Donald Scherer. For only biocentrism and ecocentrism take seriously the moral standing of living organisms that, without sentience, have contrived (or will contrive) to find ways of surviving and representing and responding to their environments, and only these normative stances can take seriously the broader tracts of space and time that are increasingly being recognized, not least because of anthropogenic changes such as climate change, as comprising the enlarged scope of ethics.

Biocentrism and ecocentrism

It is appropriate to explain at this stage that biocentrism rather than ecocentrism comprises the kind of normative stance that I consider suited to the enlarged spatial and temporal scope of ethics mentioned earlier. Certainly the kind of ecocentrism that ascribes moral standing or independent value to species and ecosystems but not to individual creatures (as, for example, in Aldo Leopold's *A Sand County Almanac* [Leopold 1949] and in J. Baird Callicott's "Animal Liberation: A Triangular Affair" [Callicott 1980, 1989]) seems to fall foul of some of the same problems as anthropocentrism, for it is unable to explain the wrongness of cruelty towards individual creatures or of neglect of the creatures in one's charge.

This said, everyone needs to take into account the interdependencies between fellow-members of the same species and between members of different species that interact with one another and thus the importance of species and of ecosystems. Biocentrists, as I shall attempt to explain in the following, can take these considerations fully into account, whereas sentientists, through omitting from consideration the great majority of species, seem unable to do this.

But what of the kind of ecocentrism that recognizes the moral standing and the independent value of individual creatures and that of species and ecosystems as well? To explain why I am unconvinced by this kind of amalgamated stance, I need to consider species and ecosystems separately.

Species can be understood either as abstractions or as transgenerational populations or as historical lineages. But when species are considered as abstractions or as types, there is no more case for adding them to the range of entities with moral standing than there is for including other abstractions in this range, such as diversity as such, or sustainability as such or other types such as rocks. Abstractions do not have a good of their own, or the ability to flourish, any more than types do, and it is instead concrete particulars with suitable characteristics that we should take into account.

I turn next to species as transgenerational populations. Here, I contend that biocentrism takes such populations into account to the full extent that this should be done. With the exception of people of the past to whom promises were made or whose wills we may be obliged to implement, there is nothing that we can do for creatures of the past, and they should not in general be taken into account (although some should be commemorated). But we should take into account both current populations and populations of the foreseeable future, and this is what biocentrism is fully able (and best advised) to do. For some purposes, this includes all foreseeable future members of certain species, such as those genetically engineered, as in their cases current actions (such as acts of genetic manipulation) will affect each and every one of these future populations; and this concern for foreseeable future members of species is entirely consistent with biocentrism.

That leaves species as historical lineages, a phrase of Holmes Rolston (Rolston 1988, 135). Rolston adds that "a species is a living historical form" and adds his support for the view of David Hull that species are historical individuals.[13] But forms, as understood by Plato by Aristotle and by their followers, are not individuals, and so the phrase "living historical form" simply produces conceptual confusion, ascribing, as it does, properties of individuals to forms or classes of individuals. Historical lineages (just like families) should rather be regarded as individuals. (But Rolston does not believe that families have intrinsic value independently of their members [Rolston 1988, 135].) Let us accept this view of species for present purposes, at least for the sake of argument, and see where it leads.

Rolston holds that species should be preserved and that there are duties to preserve them, because they each defend their own form of life. But it is the individual members of species that jointly defend their own form of life and are enabled to flourish accordingly. There is no need to posit lineages as having independent intrinsic value, over and above that of individual species members, or as defending the form of life of the lineage additionally to the defending of such patterns performed by the members. Indeed to posit this is to embark on double-counting, for it purports to find intrinsic value in the lineage additional to the intrinsic value that the individual members carry. There is, then, no basis to regard species as having independent intrinsic value, whether they are regarded as abstractions, populations, or historical lineages.

Nor is this an atomistic view. Biocentrism is not required to be atomistic in the sense of considering the good of individuals and ignoring the good of collections of them. For example, if global warming requires an entire species to migrate to a new habitat, biocentrists can take this fact into account when considering the good of each individual member; species can be important, without carrying intrinsic value. So I conclude that biocentrists can take species into account to the full extent that they should be taken into account.

I now turn to ecosystems. Once again, biocentrists can take into account the needs of individual organisms for suitable habitats and for the presence of members of the other species with which they need to interact, not least as predators or as prey. Indeed, awareness of such considerations already involves taking into account important ecosystemic factors. But before it can be claimed that ecosystems count in their own right, some principles of individuation must be available for what counts as an ecosystem and what counts as the same one over time. Yet it is increasingly being recognized that ecosystems are constantly fluctuating and that there are no principles for individuating them.[14] Hence it is unclear what would be meant by adding ecosystems to biocentrism's account of the bearers of moral standing.

If, however, ecosystems are understood as their component populations at given times, then adding ecosystems would simply (once gain) involve double-counting, since all these individuals are taken into account by biocentrism already and also by the kind of ecocentrism which includes biocentrism. Biocentrists should be open to systemic thinking, but do not need to supplement their understanding of moral standing beyond the individual creatures of the present and the future.

This said, the forms of ecocentrism that incorporate biocentrism, such as that of Rolston (see Rolston 1988), are likely on most occasions to arrive at defensible practical judgments, and their adherents can be considered allies by biocentrists in matters of responding to climate change and changing human practices accordingly. Double-counting aside, seldom will judgments based on the supposed independent value of ecosystems diverge from well-informed biocentric judgments.[15]

Moral motivation

There is a problem for every normative theory of motivating even its adherents to act accordingly, and biocentrism is no exception. Some philosophers even claim that the apparent inability of people to act in accordance with their principles counts against those principles, arguing that when people are unable to act on a principle, that is enough to show the principle to be misconceived (because "ought" either implies or presupposes "can"). But, whether or not this reasoning is valid, there are countless cases of people expressing their concern for members of other species, present and future, in practical action; hence the claim that human agents cannot act on biocentric principles miscarries. So the problem is not to be solved by discarding biocentrism but by mobilizing its resources so that it is put into practice more widely.

The case for taking climate change seriously is far stronger, given biocentric principles, than it is on the basis of anthropocentric ones and even than on the basis of sentientist ones. Thus where biocentrism is granted, people can be motivated by concern about the creatures of the ocean depths and the creatures beneath the Antarctic ice shield, whether sentient or not, and also about the entire stretch of the future on which human actions can make a difference, as well as by concern for contemporary neighbors and fellow sentient beings. (For the role of wonder in environmental motivation, see Attfield 2017).[16] Besides, these motivational resources of biocentrism link with differential possibilities for moral education to make an impact on people's behavior.

One of the best ways to foster environmental commitment is through environmental education, including love of natural outdoor places. But the relevance of such education is only indirect for people whose principles are either anthropocentric or sentientist. For such people to be motivated by visits to (say) a heath or a woodland, they need to satisfy themselves that preserving it prevents harm either to humans or to sentient creatures, and there will often be little evidence that this is the case. But it is much easier to satisfy oneself that damage to a heath or a woodland or a wetland is likely to undermine the trees, bushes, mosses, algae, and invertebrates that are its inhabitants.

Granted teaching about the ecological impacts on climate change of, for example, the erosion of heaths or woodlands or wetlands, or the impacts (both local and global) of burning vegetation, it is much easier for those with a biocentric outlook to adopt a committed preservationist attitude. Educational visits, then, combined with teaching about impacts, have a central place in fostering motivation to apply one's principles to practical decision-making and lifestyle choices. Related television programs and films can serve to strengthen such motivations.

Thus, the prospects for the development of motivations for the implementation of biocentric principles are not beset by the difficulties that might be imagined and are ampler than might be feared. Since, further, much greater motivation than is currently in evidence is clearly needed to induce people, corporations, and countries to adjust their behavior and practices in the light of climate change, there is every reason to persevere with making the case for biocentrism, as well as with environmental education. If the full resources of biocentrism are pressed into service, there may even be some chance that humanity will respond to the immense challenges posed by climate change in time.[17]

Notes

1 Cicero, *De Finibus*, 3.64; cited in Stephen R. L. Clark, "Environmental Ethics," *Companion Encyclopedia of Theology*, eds. Peter Byrne and Leslie Houlden (London and New York: Routledge, 1995).
2 Hans Jonas, *The Imperative of Responsibility: In Search of an Ethics for the Technological Age*, translated by Hans Jonas and David Herr (Chicago and London: University of Chicago Press, 1984); originally published as *Das Prinzip Verantwortung: Versuch einer Ethic für die technologische Zivilisation* (Frankfurt am Main: Insel Verlag, 1979); *Macht oder Ohnmacht der Subjecktivität? Das Leib-Seele-Problem im Vorfeld des Prinzips Verantwortung* (Frankfurt am Main: Insel Verlag, 1981).
3 This paraphrase has been borrowed from Robin Attfield, "Mediated Responsibilities, Global Warming and the Scope of Ethics," *Journal of Social Philosophy* 40, no. 2 (2009): 225–36, p. 225.
4 Kenneth Goodpaster, "On Being Morally Considerable," *Journal of Philosophy* 75 (1978): 308–25.
5 See Douglas Fox, "Discovery: Fish Live Beneath Antarctic," *Scientific American*, 2015, www.scientificamerican.com/article/discovery-fish-live-beneath-antarctica/ (accessed April 1, 2019); Robin Attfield, *The Ethics of the Global Environment*, 2nd ed. (Edinburgh: Edinburgh University Press, 2015), 171.
6 See Robin Attfield, *Value, Obligation and Meta-Ethics* (Amsterdam and Atlanta, GA: Rodopi, 1995, re-issued Leiden: Brill, 2018), 128–31.
7 Mark Sagoff, "On Preserving the Natural Environment," *Yale Law Journal* 84 (1974), 205–67.
8 Robin Attfield, *The Ethics of Environmental Concern*, 2nd ed. (Athens, GA, and London: University of Georgia Press, 1991), 148.
9 On compartmentalization, see Rebekah Humphreys, "Moral Feelings, Compartmentalisation and Desensitisation in the Practice of Animal Experimentation," in *Animals and Business Ethics*, ed. Natalie Thomas (to be published by Springer, forthcoming).

10 I have discussed the merits and demerits of anthropocentrism further in Robin Attfield, "Beyond Anthropocentrism," in *Philosophy and the Environment*, ed. Anthony O'Hear (Royal Institute of Philosophy Supplement 69) (Cambridge: Cambridge University Press), 2011, 29–46.

11 See Donald Scherer, "Anthropocentrism, Atomism, and Environmental Ethics," in *Ethics and the Environment*, eds. Donald Scherer and Thomas Attig (Englewood Cliffs, NJ: Prentice-Hall, 1983), 73–81.

12 Richard Routley (later Sylvan), "Is There a Need for a New, an Environmental Ethic?" in *Proceedings of the XVth World Congress of Philosophy* (Bulgaria: Varna, 1973), 205–10, reprinted in Robin Attfield (ed.), *The Ethics of the Environment* (London and New York: Routledge, 2008), 3–12.

13 Holmes Rolston III, *Environmental Ethics: Duties to and Values in the Natural World* (Philadelphia, PA: Temple University Press, 1988); David Hull, "A Matter of Individuality," *Philosophy of Science* 45 (1978): 335–60.

14 See James Sterba, "A Biocentrist Strikes Back," *Environmental Ethics* 20, no. 4, (Winter 1998), 361–76, at p. 370.

15 Bryan Norton has argued that anthropocentrist and non-anthropocentrist value-systems (both biocentrist and ecocentrist) will invariably converge in judgments; see Bryan Norton, *Towards Unity Among Environmentalists* (New York and London: Oxford University Press, 1991). I have replied to this view at Robin Attfield, *Environmental Ethics: An Overview for the Twenty-First Century* (Cambridge and Malden, MA: Polity Press, 2014), 120–26.

16 Robin Attfield, *Wonder, Value, and God* (London and New York: Routledge, 2017).

17 See further Robin Attfield, *Environmental Ethics: A Very Short Introduction* (Oxford: Oxford University Press, 2018).

References

Attfield, Robin. "Beyond Anthropocentrism." In *Philosophy and the Environment*, Royal Institute of Philosophy Supplement 69, edited by Anthony O'Hear, 29–46. Cambridge: Cambridge University Press, 2011.

Attfield, Robin. *Environmental Ethics: An Overview for the Twenty-First Century*. Cambridge and Malden, MA: Polity Press, 2014.

Attfield, Robin. *Environmental Ethics: A Very Short Introduction*. Oxford: Oxford University Press, 2018.

Attfield, Robin, ed. *The Ethics of the Environment*. London and New York: Routledge, 2008.

Attfield, Robin. *The Ethics of the Global Environment*. 2nd ed. Edinburgh: Edinburgh University Press, 2015.

Attfield, Robin. "Mediated Responsibilities, Global Warming and the Scope of Ethics." *Journal of Social Philosophy* 40, no. 2 (2009): 225–36.

Attfield, Robin. *Value, Obligation and Meta-Ethics*. Amsterdam and Atlanta, GA: Rodopi, 1995 (re-issued Leiden: Brill, 2018).

Attfield, Robin. *Wonder, Value and God*. London and New York: Routledge, 2017.

Callicott, J. Baird. "Animal Liberation: A Triangular Affair." *Environmental Ethics* 2 (1980): 311–28.

Callicott, J. Baird. "Animal Liberation: A Triangular Affair." In *In Defense of the Land Ethic: Essays in Environmental Philosophy*, edited by J. Baird Callicott, 15–38. Albany: State University of New York, 1989.

Clark, Stephen R.L. "Environmental Ethics." In *Companion Encyclopedia of Theology*, edited by Peter Byrne and Leslie Houlden. London and New York: Routledge, 1995.

Fox, Douglas. "Discovery: Fish Live Beneath Antarctic." *Scientific American*. 2015. Accessed 1 April 2019. www.scientificamerican.com/article/discovery-fish-live-beneath-antarctica/.

Goodpaster, Kenneth E. "On Being Morally Considerable." *Journal of Philosophy* 75 (1978): 308–25.

Hull, David. "A Matter of Individuality." *Philosophy of Science* 45 (1978): 335–60.

Humphreys, Rebekah. "Moral Feelings, Compartmentalisation and Desensitisation in the Practice of Animal Experimentation." In *Animals and Business Ethics*, edited by Natalie Thomas. To be published by Springer, forthcoming.

Jonas, Hans. *Das Prinzip Verantwortung: Versuch einer Ethic für die technologische Zivilisation*. Frankfurt am Main: Insel Verlag, 1979.

Jonas, Hans. *The Imperative of Responsibility: In Search of an Ethics for the Technological Age*, translated by Hans Jonas and David Herr. Chicago and London: University of Chicago Press, 1984.

Jonas, Hans. *Macht oder Ohnmacht der Subjecktivität? Das Leib-Seele-Problem im Vorfeld des Prinzips Verantwortung*. Frankfurt am Main: Insel Verlag, 1981.

Leopold, Aldo. *A Sand County Almanac and Sketches Here and There*. New York: Oxford University Press, 1949.

Norton, Bryan. *Towards Unity Among Environmentalists*. New York and London: Oxford University Press, 1991.

Rolston, Holmes, III. *Environmental Ethics: Duties to and Values in the Natural World*. Philadelphia, PA: Temple University Press, 1988.

Routley (later Sylvan), Richard. "Is There a Need for a New, an Environmental Ethic?' In *The Ethics of the Environment*, edited by Robin Attfield, 3–12. London and New York: Routledge, 2008.

Routley (later Sylvan), Richard. "Is There a Need for a New, an Environmental Ethic?' In *Proceedings of the XVth World Congress of Philosophy*, 205–10. Bulgaria: Varna, 1973.

Sagoff, Mark. "On Preserving the Natural Environment." *Yale Law Journal*, 84 (1974): 205–67.

Scherer, Donald. "Anthropocentrism, Atomism and Environmental Ethics." In *Ethics and the Environment*, edited by Donald Scherer and Thomas Attig, 73–81. Englewood Cliffs, NJ: Prentice-Hall, 1983.

Sterba, James. "A Biocentrist Strikes Back." *Environmental Ethics* 20, no. 4 (Winter 1998): 361–76.

6 Evaluating climate change with the language of the forms of life

Claudio Campagna and Daniel Guevara

When air conditioning was invented, in the early 1900s, it was referred to as "manufacturing weather" (see Revkin 2018). We often call air conditioning appliances "climatizers." By definition, climate is weather but on a scale of decades – today's rain is weather, the average rain in decades is climate. The idea that weather may be manufactured is pretentious, yet some benefits of weather control would be handy in some circumstances. A century ago, "weather alchemists," such as Fleming (2010) calls them, confronted drought with explosives. The idea was "to attack the atmosphere on multiple fronts with dynamite and blasting powder, mortar shells, smoke bombs, electrified kites, exploding oxy-hydrogen balloons, and even fireworks" (Fleming 2010). This author cites the May 1923 issue of *Popular Science Monthly*:

> Think of it! Rain when you want it. Sunshine when you want it. Los Angeles weather in Pittsburgh and April showers for the arid deserts of the West. Man in control of the heavens—to turn them on or shut them off as he wishes.

"Man in control of the heavens" calls to mind the geoengineering of climate. Indeed, by facing anthropogenic global warming as a *fait accompli*, geoengineering "solutions" turn into pragmatic responses for mitigating disastrous consequences at a planetary scale. According to this picture, climate disruption through the burning of fossil fuels works like an unintended consequence of "weather" manufacturing. Then, the natural thought of the techno-fixers is to take responsibility and mend the harm.

While techno-fixers and supporters of planetary stewardship see opportunity, conservationists agonize over the effects of climate change on life and habitats. The Earth's ecosystems are already highly fragmented and degraded. Climate change will make things worse, and many species will go extinct. These gloomy predictions may be avoidable, but that depends on our will to act and do the right things. This raises ethical questions about why humanity does not seem primarily concerned with the life forms we affect. In evaluating the disastrous consequences of climate change and proposing interventions to mitigate them, the primary concern seems to be focused on the urgencies of human safety and welfare, including the future of humanity and its economy. We here face the ancient problem of practical reason: what ends we should pursue and what we should treat as ends in themselves.

The fact that today most everyone in world-class conservation thinks of practical reasoning primarily in instrumental, homocentric terms (e.g., reasoning about the impact of climate change on human welfare) results, we here argue, from important limitations in our normative language. Instrumentalist language plays such an important role largely because of a lack of alternative models to the widely accepted, homocentric, instrumental reasoning; it is more natural to us to represent states of affairs under values that reflect back on our needs than on the needs of other species.

So language plays a role in the difficulties we are confronting in trying to evaluate and redress the consequences of climate change; among other things, it installs a bias towards our own good.

The word *language* itself can be used in various ways. It can informally refer to a roughly defined domain of discourse, as in the "language of conservation" or the "language of climate change" or the "language of mathematics" or the "language of poetry." Reference here is made to the terms, phrases, metaphors, and so on characteristic of the domain. But "language" may also refer to "grammar," where that does not concern superficial issues, like whether a final "ed" or a "t" forms the past tense of a verb but rather where it concerns the logical structure of language, which makes contents intelligible and makes possible the expression of thought generally. We will here be concerned with ". . . both senses of language . . ." but with emphasis on the latter. It is only by attending to the logic and grammar of "life forms" that we can break through the limitations of the dominant language and thinking in conservation and uncover the full implications of understanding what it is to be a living thing.

Thompson (2004, 2008) shows that we only understand something as a living thing insofar as we understand it as instantiating a form, like that represented by the shape of a life cycle. Most importantly for our purposes, the concept of a life form gives us the primary normative standard for evaluating whether a creature is doing well or badly or whether it is harmed or benefited. This is because the form of its life serves as a template for judging whether the creature is alive or not, healthy or sick, flourishing or dying, defective or sound, in the first place. This is what Foot (2001) calls "natural goodness (or badness)." The work of Thompson and Foot provides a fresh start on questions like the following: If it is wrong, what kind of wrong is an extinction event caused by climate change originating in human agency? It does so by making the forms of life the primary standard of good or bad, harm or benefit, and so on.

The language we now commonly rely on in describing and assessing the effects of climate change obscures primary goodness in this sense. The typical evaluation of the effects of climate change begins, of course, with empirical understanding of facts. These facts, regarding, for example, economic and social effects, then lead to familiar ethical judgments related to human welfare. All manner of standards are appealed to in this context, mostly human centered. In contrast, the language of "life forms" gives us a standard set, in the first place, by the forms of life. As we explain in more detail in the following, this eliminates the "fact-value" dichotomy and has the potential to guide us ethically through central concern with the forms of life generally, rather than just human life. We recognize the consequences of human-induced climate changes to our own form of life. From the point of view of natural goodness, as the primary standard, it is irrational or at least unreasonable to favor the good of just one form of life (our own), while marginalizing the others and their representatives.

Eileen Crist (2007) pointed out, a decade ago, the bias towards the human in discussions of climate change:

> Instead of confronting the forms of social organization that are causing the climate crisis . . . climate-change literature often focuses on how global warming is endangering *the culprit*, and agonizes over what technological means can *save it* from impending tipping points.
>
> (Crist 2007, 35, italics in the original; see also Fernandez 2020)

Likewise, Crist points to a discourse that enhances techno-solutions but that does not address the crisis as primarily an ecological one: "climate-change discourse encourages the continuous marginalization of the biodiversity crisis . . . which despite decades of scientific and environmentalist pleas remains a virtual non-topic in society, the mass media, and humanistic and other academic literatures" (Crist 2007, 36). If "confronting the forms of social organization that are causing the climate crisis" is what we need, we will argue that it will never happen unless we are aware of the workings of the dominant discourse and its departure from a language that centers on the forms of life.

"Life forms" and "species" are used almost interchangeably by Foot (2001) and Thompson (2004, 2008), but we must be careful to distinguish them in light of how conservationists use "species." These examples, drawn from the 2019 IUCN Red List page on polar bears (www.iucnredlist.org/species/22823/14871490), illustrate first typical life form and then species language:

> In late autumn, pregnant polar bear females enter dens in snow drifts or slopes on land, close to the sea or on sea ice, where they give birth a few months later.
>
> Loss of Arctic sea ice due to climate change is the most serious threat to Polar Bears throughout their circumpolar range. . . . There is a potential for large reductions in the global Polar Bear population if sea-ice loss continues.

In the first sentence, we have life-form language, expressed by natural historical descriptions of the specific shape or cycle of the form. But the rest reflects "species" language as commonly used in biology. The use of "species" in the context of climate change is mostly intended in the way conservation biologists typically use it: where what is meant is "populations" of individuals grouped by certain empirically determinable criteria of categorization (see Nolt 2020) and described according to statistical parameters. This "population language" is useful when trends in numbers or changes in distribution help promote interventions to protect the diversity and functions of living things. But if we are concerned about the language that evaluates the impact of anthropogenic alterations of climate on life in general, it is the language of "forms of life" that is most helpful. It represents life as a fundamental logical category and also provides a standard for the normative evaluation of representatives (Thompson 2008); for example, we know something is wrong, simply on understanding the life form, if we find a polar bear far from the sea in late autumn.

After considering some background notions critical for further analysis, we will address the theory of "natural goodness." Our own contribution to the theory lies in the application to questions of conservation. Conservation practice tends to focus on measuring the ecological impact of climate change, employing a mixture of terms and concepts, including the data and predictions of the relevant empirical sciences (e.g., Webster et al. 2017). There is usually also some attempt, implicit at least, to make normative assessments, though mainly concerned with the effects on human welfare. What Foot and Thompson have provided us is a structure or framework for bringing life forms to the center of conservationist concerns and in a way that makes the life forms themselves the standard for the evaluations of the effects of our threats to them.

Why language?

It goes this way: (i) Humans have become a geological force, leaving a footprint equivalent to major geological processes at planetary levels. (ii) That state of affairs is captured in the concept of the Anthropocene: "the new human-dominated geological epoch" (Lewis and Maslin 2015; Zalasiewicz et al. 2010). (iii) The Anthropocene has introduced to the world an apocalyptic threat: "The ultimate drivers of the Anthropocene . . . if they continue unabated through this century, may well threaten the viability of contemporary civilization and perhaps even the future existence of *Homo sapiens*" (Steffen et al. 2011). (iv) Yet, these premonitory warnings have not had the results we might have expected. In a summary of her book *In Catastrophic Times*, and referring to climate change, Isabelle Stengers (2015) writes, "Our governments are totally incapable of dealing with the situation. Economic warfare obliges them to stick to the goal of irresponsible, even criminal, economic growth, whatever the cost." (v) Humans are indisputably affected, but so also are many other forms of life. In *Defaunation in the Anthropocene* (Dirzo et al. 2014), the authors argued that a widespread decrease in population sizes across taxa (defaunation) "is both a pervasive component of the planet's sixth mass extinction and also a major driver of global ecological change." It is predicted that "one-sixth of all species may go extinct if we follow 'business as usual' trajectories of carbon emissions" (Lambers 2015 based on Urban 2015). (vi) Climate change, just one phenomenon associated with the Anthropocene, has been the most effective at forcing us to think through the pros and cons of a human-dominated world (Steffen et al. 2011).

(vii) Today, climate change and the Anthropocene dominate the discussion of environmental concerns.

A notable consequence of the preceding state of affairs is that the focus on climate change, as the central environmental problem, may be displacing the attention of the public to the biodiversity crisis. While media coverage of climate change issues expands, the crisis of extinctions receives less coverage (Legagneux et al. 2018). The drama of this scenario is that this "marginalization of the biodiversity crisis" (Crist 2007), due to an extreme human-bias in our judgment, reinforces the bias via the logic and grammar of the language that evaluates what the crisis is about.

Crist (2007), applying insights from the Wittgensteinian philosopher Peter Winch, makes this same point in a link between the Anthropocene, climate change, and language: "We cannot be so naïve as to dissemble that to speak of the Anthropocene is merely to describe, because, in fact, it is also to act: such speech anchors it and participates in its consolidation" and

> To propose the "Anthropocene" as a description of reality . . . is to rescind responsibility for the way the proposed concept, in turn, acts upon the very reality it purports to merely describe: reinforcing it, sharpening its contours, and, through the extraordinary power of language to mold the world into experience and meaning, ultimately legitimizing it. In brief, proposing a concept of this magnitude does not simply reflect a state of affairs, but also amounts to crystallizing and affirming that state of affairs.

In our view, Crist (2007) correctly identifies the way the language of climate change reinforces evaluative discourses detached from the effects on life forms. The simple use of the phrase "climate change" engages with other logically compatible terms and phrases, in a way analogous to gears in the transmission of a car. The structure of the language favors certain evaluative terms but not others, gathering momentum and reinforcing a direction in our thinking. Under this momentum, the threat to non-human forms of life does not easily find a place in our thinking. The logical tendency is instead to focus on geoengineering and the efforts to manipulate and get control of the physics and chemistry of climate. The prevailing tendency is to think that the same science and technology that drove us to industrial development and consumption of fossil fuels can also be developed to mitigate the undesirable consequence of living in the Anthropocene. Being aware of how this blinds us to the value of life in all its great diversity of forms, as Crist is urging, is a necessary step to breaking the cycle and redirecting our concern to the life forms themselves.

The common discourse of climate change

Typical language related to climate change involves the language of facts, socioeconomic costs, and interventions for mitigating negative human consequences, among other things. The very title of a report of the IPCC (2018) illustrates the language:

> Global warming of 1.5°C. An IPCC Special Report on the impacts of global warming of 1.5°C above pre-industrial levels and related global greenhouse gas emission pathways, in the context of strengthening the global response to the threat of climate change, sustainable development, and efforts to eradicate poverty.

This language is in line with evaluations that see nature instrumentally, we find it even in statements of facts that are apparently aimed at making us aware of the plight of non-human species:

> Global warming of 1.5°C is projected to shift the ranges of many marine species to higher latitudes as well as increase the amount of damage to many ecosystems. It is also expected to drive the loss of coastal resources and reduce the productivity of fisheries and aquaculture.

Life and its diversity are not a primary concern: "Impacts of climate change in the ocean are increasing risks to fisheries and aquaculture via impacts on the physiology, survivorship, habitat, reproduction, disease incidence, and risk of invasive species."

The IPCC informs decision makers; its focus on humans is thus expected. Yet, in reviewing scientific evidence and placing it in the context of development, the discourse of the IPCC installs a way of talking and thinking whose logic and grammar have far-reaching consequences beyond decision and policy making, further expanded by media, education programs, funding agencies, research institutions, and, obviously, governments.

The language of the IPCC illustrates modes of expression, each with its own rules, that guide an assembly of words, concepts, phrases, and so on around empirical facts and their interpretation. It is in these rules that we are particularly interested. Using language is an action shaped by rules, a point which led philosopher Ludwig Wittgenstein (1958) to compare language to games. The analogy helps us keep in mind that language use necessarily involves actions with consequences. Crist (2007) makes the point that even in supposedly "purely" descriptive discourse about climate change, the "moves" we make in the language game are actions that could actually make the problem worse (see Campagna, Guevara, and Le Boeuf 2017). The language games suited to the purpose of evaluating climate change in the context of the Anthropocene, with the needs of governments in mind, follow rules that make it difficult to articulate primary concern with non-human forms of life.

It is important to emphasize in this context that conservation is the practice of a normative philosophy—including an ethical philosophy; thus normative evaluations are unavoidable and to be expected. Conservationists are not typically troubled by the plurality of evaluative language, as all language games (e.g., of sustainability or intrinsic or aesthetic value) seem to work toward a common, desirable goal. Yet, certain discourses drive the terms and standards in our evaluation of climate change, eliminating from view the life forms themselves. We need a new standard of evaluation, which in turn requires a new logic, new rules that guide moves in a new game that captures with its logic the centrality of the notion of a form of life.

The language game that demonstrates a normative standard in the logic of forms of life

The grammar and logic of "life form" terms themselves give us the normative standard that, we suggest, is the standard most relevant to conservation in general, including assessment of the effects of climate change on species. This logic and grammar is brought out by what Thompson calls Aristotelian categoricals (ACs). This is an example:

> flying squirrels have their limbs and even the base of the tail united by a broad expanse of skin.
>
> (Darwin 2001, 165)

These types of "vital descriptions" (Thompson 2004) have distinctive and interesting properties. The first thing to see is that the subject term, *flying squirrel*, does not refer to a particular individual or group of individuals (e.g., a population), since an individual squirrel (or even most or all of a population) might be injured or may be malformed and the categorical nonetheless be true. ACs are generics. Thus, life forms and their individual bearers are logically distinct but mutually interdependent: the form functions as a template and standard for identifying the bearers of the form and for evaluating whether they are doing well or badly, harmed or benefited, and so on, in terms of the ACs true of them. Foot argued that those categoricals that describe a form of life (the necessities for its survival, development, or reproduction) not only represent the empirical description of the form but serve to evaluate it normatively. The role of the categorical is shown in propositions such as

> flying squirrels have their limbs and even the base of the tail united by a broad expanse of skin, *which serves as* a parachute and allows them to glide through the air to an astonishing distance from tree to tree.
>
> (Darwin 2001, 165, emphasis added)

Darwin's sentence is typical of the teleological (functional or purposive) language used in natural historical judgments in which "it serves to" is often represented by the conjunction "in order to." We may extend our knowledge of specific life forms by indefinitely linking all the true natural historical judgments in this teleological way. The teleology illustrated in Darwin's sentence is not that of the purposes of an intelligent designer. The ACs represent facts but also give us the natural standard according to which we can judge whether bearers of the kind are doing well or not. If we know the basics about flying squirrels, we know, for example, that a representative may not be doing well if it cannot glide. We have reason to look for a defect due to injury or disease, for example. These are judgments about NG and NB, and they are applications of the standard of evaluation given by the template, the ACs, of the life form. These naturalistic objective judgments apply to all forms of life in the same generic way, differences emerge as the specific kinds of life take their specific shapes, with their indefinitely long list of ACs describing their specific forms.

We follow Foot in thinking that this is the primary standard of evaluation, a generic standard based in the very idea of life. Ethical evaluations are just based in life-form specific ACs, those for humans or creatures like us, with rational agency and the need to cooperate through mutual benefit and trust and so on. The great majority of life forms cannot then be judged as inferior or as lacking a quality by not having an AC that requires fulfilling needs of the moral kind, as those needs are ACs of the human life form, the same way that having a stinger is part of the goodness of bees (Foot 2001).

We propose that whatever values or standards are now commonly in play in conservation, influencing its practice, ought to be organized around the standard of NG, if what we are primarily concerned with is the living things and the forms of life they bear. Our evaluation of representatives of a form of life, for example, the polar bear, will then be grounded in what is judged good or bad, according to the standard set by the ACs of polar bears. This standard is more fundamental than sentience or pleasure and pain or welfare or cognitive ability or any other of the values typically appealed to by conservationists (see Nolt 2020). And it is not limited to the other animals (see Fernandez 2020) but extends to all forms of life and the organisms that bear them.

We can see more clearly now how the language of forms of life is logically different in kind, and more basic, than the typical biological concept of a species. Biology builds the concept of a species up from individual living things to the level of the populations, grouped according to various properties they have in common. But this overlooks Thompson's fundamental insight that it is impossible to identify some individual thing here and now as a living thing, unless we can understand it as instantiating a stage of its life form, its life cycle. We know that this thing here and now is a caterpillar only because we understand that it is the sort of thing that becomes a butterfly—or that something is wrong if it does not.

The language game of life forms, by articulating the indefinitely long chains linking up the ACs proper to the forms, reveals the standard that directly bears on the question of the well-being of living things and what is necessary to the conservation of the forms of life they bear.

The practical use of Thompson's and Foot's insights

NG ought to be the primary standard of evaluation for conservation, for it is a standard directly and objectively given by the concept of life itself. It is not the only standard: there is the aesthetic value of nature, its instrumental value to human welfare, its sacredness in religious terms, or even its "intrinsic value" in the many senses of that term. But what we have in NG that we do not have in these others is a clearly defined, objective, and primary standard, directly relevant to the life forms themselves and to the well-being of the creatures that bear those forms for reasons that are stated in the language of the teleological, natural historical statements (like Darwin's) that make up the ACs.

We can only begin to sketch the practical implications of a commitment to NG as the basis for our normative evaluations in conservation; its grammar and logic is clear enough, we believe, but given the dominant language and thinking in conservation, it is a challenge to master the new language game in all its implications. The language game for assessing the effects of climate change on non-human living things is difficult to think through, because the logic of "species-as-populations" exerts an influence: it works as if there were a predisposition to understand life as fulfilling criteria that could be applied independently of the concept of life forms.

The biological concept of species grounded in population thinking obscures NG. It places at the center of our evaluations trends in numbers and shifts in distribution that are not the language of ACs. Inasmuch as they are ways of grouping living things, the various uses of "species" in biology presuppose the idea of a life form, but this goes unnoticed because what explicitly concerns biologists in their taxonomies according to "populations" is often not pertinent to this more fundamental and essential concept of form. Standard criteria used to evaluate good and bad regarding a species are not necessarily guided by the logic of ACs. Species are generally described following taxonomic rules useful to the particular science involved, without a required appeal to first principles. So, for example, a spot in the plumage of a bird may be judged important (or convenient) for taxonomical purposes, before finding an AC that expresses the necessary role of the spot as determined by the form of life (Foot 2001). No AC has the form "S should not go below X (numbers) for the species to be viable," or "The Ss should not decrease their distribution by more than X percent in order to be doing well." These population-related statements may be traced back to ACs, but the logic of the language must be changed to "The S gives birth in a den to twin cubs," or "Ss live in ice-covered waters" or like descriptions given by the natural history of the kind.

If conservation is to accept NG as the primary concept of goodness, it will make its primary standard of good and bad, right and wrong, and so on, the standard set by the forms of life and the ACs that make our groupings of living things into populations possible in the first place. That a population is expanding may be judged good by some measures but not on NG if the expansion is due simply to the increase of individuals that are poor representatives of the relevant form of life. Such representatives may survive, if they are lucky, by making, for example, use of resources that are the by-products of the human form of life (e.g., the food we leave behind in national parks), unsupported by any ACs. Or a species may be "extinct in the wild," according to a category of the Red List, which means that individuals survive *ex situ*; hardly any of the goods that such representatives might possess or enjoy (considerable comfort and affection from zookeepers, say) has anything to do with their NG. Finally, specimens may be the product of synthetic biology, thus lacking any primary goodness: being essentially artifacts or like domesticated animals.

Our argument is not against *ex-situ* conservation; we only point out that these organisms cannot be evaluated as doing well, if we use the standard set by forms of life themselves. We are raising the question of what it is that conservation is supposed to be conserving, or primarily concerned with, if not the values that relate directly to the forms of life themselves. Focused on biological species as populations, we protect diversity and ecological function and aim at sustaining demographic processes or maintaining community interactions, if possible by facilitating processes dependent on natural selection, acclimatization, and ecological reorganization (Webster et al. 2017). We may even assist populations by facilitating colonization and migration. But unless these interventions can be shown to be guided by NG of the forms of life, we must ask what their point is.

Climate change alters conditions that may make it difficult, if not impossible, for any representative to fulfill the necessities of the form. The result may be an acceleration of the extinction rate after a period of lack of adaptation, with many representatives performing as defective. But climate change may also create pressures that resilient populations overcome through new adaptations, where it will be difficult to say whether we have a new life form emerging or not. Whatever survives under climate pressures, and their rates of change, may include representatives and forms that get a foothold through processes that work under great perturbance, independent of the primary goodness of the form, as happens with artificial selection. It may also be unclear what to think about life forms under such changes. It is in such circumstances that life forms and species-as-populations show their differences, for in relying on the latter, biologists may have no reason to raise such questions or to care about differences crucial to those judging according to ACs.

On the approach we propose, these questions about the evaluation of individuals in altered environments must be handled in the same way we have always done in representing life and life forms, that is, through natural historical judgments. The standard of NG involved is always, at a general level, the same: it is given by the ACs of the relevant life form. Under normal conditions, if we take a "snapshot" of the life forms at any point in time, as represented by individual bearers of the forms, we can begin to determine the ACs that characterize them and set the normative standards. We know how to determine this, even for the many novel creatures we discover, in part because we understand the natural processes that have resulted in any given form we are trying to understand. In contrast, radically fast rates of change that override these natural processes will for the most part probably wipe out extant life forms. Whatever manages to hang on will probably be a degraded representative of the original form: a bad but "lucky" representative of its kind, enjoying what life it has thanks to "goods" having little to do with the ability to fulfill its form of life.

What is unclear, then, is not the general standard set by the concept of a form of life but what to look for in determining the specific shape of a form, the specific

ACs, under radically perturbed conditions. It is always a challenge to determine the specific shape of a new life we encounter, but the challenge is being met all the time by natural scientists and biologists, as they encounter and describe new, often very strange, forms of life (see Thompson 2004), against the background of evolution. The questions that arise from the strangeness introduced by perturbed processes like climate change, or synthetic biology, are difficult in a different way in that they raise philosophical problems about "nature" or "natural." What NG theory holds is that life can only be understood in terms of the ACs of a given life form. What we understand thereby is "natural life," something we can understand without appeal to the products or by-products of our agency. As Thompson points out, human artifacts or machines are also characterized by "in order to" clauses that seem to follow the same grammar as ACs (think of how the explanation goes for how a carburetor works). The fundamental difference lies in the fact that artifacts instantiate functions that are not ACs. The "in order to" statements look the same, but the underlying logic is different. Artifacts instantiate an idea that their inventor had in the first place (and thus presuppose an intelligent life form). The agent's idea is the template. In contrast, natural life exists as an autonomous process, and of course, many life forms exist, or have existed, that no one has or will ever know or form any idea of.

The standard of well-being that living things reveal is independent of any ideas we might or might not have of them: a standard as objective and mind independent as the creatures themselves. At the most generic level, then, NG is an entirely invariant, objective standard grounded in the relationship of natural life form to bearer.

Schematically, conservation practice proceeds following four steps: (i) determining the facts, (ii) evaluating facts normatively, (iii) recognizing in the evaluation reasons to act, and (iv) intervening on the basis of these reasons. Thus, as noted, conservation is a normative practice. But with the language of populations, we may be evaluating facts about the effects of climate change on living things without being guided by how our values relate to the forms of life themselves, neglecting the standards or norms that should be guiding us clearly and objectively, through the relevant ACs. The language of NG ought therefore to be of great interest to the conservation practitioner because of its power to express values in terms of the necessities of the life forms themselves.

Using NG rationale, we may conclude with the following observations:

1 Climate change is a "life-form changer." It disrupts and alters life cycles by interfering with what grounds the specific content of the relevant ACs. As a result, we may no longer be able to make NG judgments about the living things that emerge. It will not be clear whether we have a new form or very defective representatives of an old form. Conversely, previously judged defective representatives may thrive, confusing our evaluation of goodness and badness.

2 Representatives that flourish under new, hard-to-identify ACs may be said to belong to the same biological species, depending on the criteria employed for

various purposes in circumscribing populations. This may include defective representatives of species. Biologists may still recognize the species as the same and describe them as threatened populations with new adaptations. For example, it may be that female polar bears manage to give birth out of the den to one cub, and on the criteria for populations, we may continue to identify them as the same as the pre-climate-change species. This is not possible with life form language. The form of life intimately depends on new ACs, and we would have to ask whether we were encountering a new life form or a defective old one. This illustrates how the notion of species, which is subject to a plurality of alternatives (de Queiroz 2007; Nolt 2020), differs from that of form of life. The impact of climate change on ACs may be either not captured by the conservation language grounded in biological species or be covered up by a language game that does not primarily relate to the life form, for example, as talk of "changing trends" and "distributions" in populations. This has consequences on our reasons to act and on what we judge to be a proper intervention.

3   Species and life form language may converge in evaluating an extinction as a bad, but the reasons may differ. The ACs of an extinct species must be stated in the past tense. There is nothing answering to the template or standard set by the form anymore, and there never will be. All NG instantiated by the bearers of that species is lost forever. In contrast, a species-as-population view may represent an extinction as a stage of population zero, or as altered processes in the trophic chain. According to this language game, it could be that the only value lost is instrumental or aesthetic or culturally determined. All secondary to the primary goodness of the form of life according to NG.

4   NG is determined through a snapshot of the form. After climate change, hangers-on will in some cases evaluate as defective, or as poor representatives. Only with time, as the "old" ACs are found unfit as standards and new ones are understood, will our judgment change templates (new ACs). The difference in snapshots will indicate a different form of life caused by climate change. The species approach, as it focuses on population variables, would tend to miss entirely the effects of change on the ACs and thus on the loss of the natural goodness of individuals.

5   As it may become harder and harder to say of the living things that hang on or emerge or adapt what it is that they do and towards what ends their various functions, processes, parts, or activities work, humans will not be exempt from this confusion. We may struggle to make sense of what is flourishing for ourselves.

6   If, under climate change, good representatives of a form of life find no chance to fulfill their ACs, species language will likely evaluate the state of affairs as "a threat." "Threatened species" is population language that is not specific about the necessities that representatives can no longer satisfy, in terms of their life forms. Conversely, in form of life language, "a threat" will be represented by a list of ACs that cannot be fulfilled anymore or that may be only defectively fulfilled. We will have then reasons to evaluate this state of affairs as a bad or wrong in terms of primary goodness.

7   In the discourse of species as populations, climate-driven processes do not have to be thought of as different from evolutionary change. The fact that some argue that the Anthropocene is forcing the "sixth mass extinction" suggests they equate us (as agents of extinction) with profound changes in the chemistry of the atmosphere, for example. Life-form language, conversely, forces us to see representatives that had once been good as harmed by our agency, and their extinction raises the possibility of ethical questions that do not arise from past mass extinctions. We must then confront our responsibility for harms done, which should motivate different reasons to act. We will thus distinguish between a natural extinction and the annihilation of species by our routine and avoidable behaviors.

8   The perturbations of climate change, if known, will be captured by life form language as a fast change in the ACs. This would work as a language indicator of perturbance. Population language will integrate all these changes in changes in population trends. Indeed, the rate of change makes climate change resemble a generalized, non-directional, gigantic "artificial" selection force—artificial in the sense of perturbance originated by human contrivances, as with selection of traits in domesticated plants and animals. Climate change may be comparable to an artificial selection force of global dimension. New natural forms of life would then share resemblances, conceptually, to the origin of domesticated species. Whatever goodness they may enjoy will be secondary goodness. Indeed, the fact that it enjoys this secondary goodness may be due to conservation interventions themselves.

Conclusions

We have discussed the use of language, and its rules, in evaluating the effects of climate change and the normative concerns indispensable to conservation. When understood in terms of life forms, we see that the issues raised by the radical effects of our routine behaviors are ethical issues inasmuch as they concern the fabric of life at its most profound levels. But it is possible that we are unprepared to revise our values. Aldo Leopold (1949) suggested long ago a limitation in capturing with language some values in nature: "Our ability to perceive quality in nature begins, as in art, with the pretty. It expands through successive stages of the beautiful to values as yet uncaptured by language." In this same vein, Bruno Latour (2014), in a conference entitled "How to Speak the Language of Gaia," and referring to crossing planetary boundaries (see Steffen et al. 2015), said,

> It is easy for us to agree that humanity is not equipped with the mental and emotional repertoire to deal with such a vast scale of events; that they have difficulty submitting to such a rapid acceleration for which, in addition, they are supposed to feel responsible.

Latour may be referring to cognitive barriers in evaluating aspects of the empirical state of affairs, which may relate to language barriers (Campagna and Guevara 2014).

With climate change, we may cross the irreversible threshold of biosphere integrity. Therefore, a discussion that breaks through the limits of the present intellectual and emotional discourse is urgently needed, despite the profound difficulties and risks that may be involved in trying to break through them. The extraordinary disruption that climate change causes makes it difficult to keep our moorings in the language that represents life. This is one of the best indicators of the threat posed to life. Being aware of the pressure that climate change exerts on a primary logical category that structures human relation to the rest of life is but a step in a promising direction. We seek a normative standard that guides us in terms of what is good or bad for life forms and the individual living things that bear those forms. The current language games must engage, or be replaced by, one that pertains to life forms centrally.

Acknowledgments

We thank the Humanities Institute and the Center for Public Philosophy, at the University of California Santa Cruz, as well as Deans Paul Koch and Tyler Stovall for supporting our Language for Conservation Project. We gratefully acknowledge the support of Synchronicity Earth and Eulabor Institute for funding the Montefortino workshop, where some of these ideas developed. We are thankful to Zack Walsh and Brian G. Henning for inviting us to contribute to this volume.

References

Campagna, C. and D. Guevara. "Conservation in No Man's Land." In *Keeping the Wild Against the Domestication of Earth*, edited by G. Wuerthner, E. Crist, and T. Butler, 55–63. San Francisco: Foundation for Deep Ecology and Washington, DC: Island Press, 2014.

Campagna, C., D. Guevara, and B.J. Le Boeuf. "De-scenting Extinction: The Promise of De-extinction May Hasten Continuing Extinctions." In *Recreating the Wild: De-extinction, Technology, and the Ethics of Conservation*, edited by G. Kaebnick, 48–53. New York: The Hastings Center, 2017.

Crist, E. "Beyond the Climate Crisis: A Critique of Climate Change Discourse." *Telos* 141 (Winter 2007): 29–55.

Darwin, C. *On the Origin of Species*. A Penn State Electronic Classics Series Publication. 2001. www.f.waseda.jp/sidoli/Darwin_Origin_Of_Species.pdf.

De Queiroz, K. "Species Concepts and Species Delimitation." *Systematic Biology* 56, no. 6 (2007): 879–86. doi:10.1080/10635150701701083.

Dirzo, R., H.S. Young, M. Galetti, G. Ceballos, N.J.B. Isaac, and B. Collen. "Defaunation in the Anthropocene." *Science* 345, no. 6195 (2014): 401–6. doi:10.1126/science.1251817.

Fernandez, L. "Climate Ethics Bridging Animal Ethics to Overcome Climate Inaction: An Approach from Strategic Visual Communication." In *Climate Change Ethics and the Non-Human World*, Routledge Research in the Anthropocene, edited by B. Henning and Z. Walsh, 33–48. London: Routledge, 2020.

Fleming, J.R. "Manufacturing the Weather: Distillations." *Science History Institute*. 2010. www.sciencehistory.org/distillations/magazine/manufacturing-the-weather.

Foot, P. *Natural Goodness*. Oxford: Clarendon Press, 2001.

IPCC. "Summary for Policymakers." In *Global Warming of 1.5°C: An IPCC Special Report*, edited by V. Masson-Delmotte et al., 32 pp. Geneva, Switzerland: World Meteorological Organization, 2018. https://report.ipcc.ch/sr15/pdf/sr15_spm_final.pdf.

Lambers, J.H.R. "Extinction Risks from Climate Change." *Science* 348, no. 6234 (2015): 501–2. doi:10.1126/science.aab2057.

Latour, B. *How to Speak the Language of Gaia*. Anthropocene Lecture. Berlin: HKW, 2014. www.bruno-latour.fr/lectures.

Legagneux, P., et al. "Our House Is Burning: Discrepancy in Climate Change vs. Biodiversity Coverage in the Media as Compared to Scientific Literature." *Frontiers in Ecology and Evolution* 5 (2018): 175. doi:10.3389/fevo.2017.00175.

Leopold, A. *A Sand County Almanac and Sketches Here and There*. New York: Oxford University Press, 1949.

Lewis, S.L. and M.A. Maslin. "Defining the Anthropocene." *Nature* 519, no. 7542 (2015): 171–80. doi:10.1038/nature14258.

Nolt, J. "Climate Change and the Loss of Nonhuman Welfare." In *Climate Change Ethics and the Non-Human World*, Routledge Research in the Anthropocene, edited by B. Henning and Z. Walsh, 10–22. London: Routledge, 2020.

Revkin, A. *Weather: An Illustrated History: From Cloud Atlases to Climate Change*. New York: Sterling Publishing Co., 2018.

Steffen, W., J. Grinevald, P. Crutzen, and J. McNeill. "The Anthropocene: Conceptual and Historical Perspectives." *Philosophical Transactions of the Royal Society* 369 (2011): 842–67. doi:10.1098/rsta.2010.0327.

Steffen, W., et al. "Planetary Boundaries: Guiding Human Development on a Changing Planet." *Science* 347, no. 6223 (2015): 1259855. doi:10.1126/science.1259855.

Stengers, I. *In Catastrophic Times: Resisting the Coming Barbarism*. London: Open Humanities Press, 2015.

Thompson, M. "Apprehending Human Form." *Royal Institute of Philosophy Supplement* 54 (2004): 47–74. doi:10.1017/S1358246100008444.

Thompson, M. *Life and Action: Elementary Structures of Practice and Practical Thought*. Cambridge, MA: Harvard University Press, 2008.

Urban, M. "Accelerating Extinction Risk from Climate Change." *Science* 348, no. 6234 (2015): 571–73. doi:10.1126/science.aaa4984.

Webster, M.S., M.A. Colton, E.S. Darling, J. Armstrong, M.L. Pinsky, N. Knowlton, and D.E. Schindler. "Who Should Pick the Winners of Climate Change?" *Trends in Ecology & Evolution* 32, no. 3 (2017): 167–73. doi:10.1016/j.tree.2016.12.007.

Wittgenstein, L. *Philosophical Investigations*. New York: Wiley-Blackwell, 1958.

Zalasiewicz, J., M. Williams, W. Steffen, and P.J. Crutzen. "The New World of the Anthropocene." *Environmental Science and Technology* 44, no. 7 (2010): 2228–31. doi:10.1021/es903118j.

7 Thinking through the Anthropocene

Educating for a planetary community

Whitney A. Bauman

The epistemologically imagined "divides" in Western modern thought between religion and science, the humanities and the natural sciences, and humans and nature develop from a long series of conceptual technologies that separate faith and values from reason and facts, subjective experiences from objective reality, and human beings from the rest of the natural world. Monotheistic theologies of human exceptionalism,[1] the Cartesian individual, the Lockean understanding of property, and the technico-scientific understanding of nature as resource are all examples of these technologies that both rely on the epistemological divide and reinforce that divide through policing bodies and ideas within the planetary community.[2] This "secular" space that creates these divides is the result of a long process by which the sciences become more and more focused on the reductive-productive materialistic model of knowledge production, while religion and philosophy get pushed away to transcendent or universalistic understandings of knowledge. Western modernity is this very secular space between reductive materialism and universal, transcendent truths in which both "nature" and "god" are dead, leaving humans as the exception to the rest of the natural world. This divide culminates, in other words, in the contemporary concept of the Anthropocene, which recapitulates the idea of human exceptionalism and to a great degree vacates agency from the rest of the natural world.[3]

Unfortunately, the disciplinary divisions within contemporary Western universities are both a product of this "great divide" and producers of the tools that help to reinforce it. In addition, these divides help to create the grounds from which our ethical thinking emerges: they are part of the formal structure that shape the character, values, and ideas of who we (humans) are and what we (humans) ought to do ethically speaking. In order to get beyond an anthropocentric ethic, then, Moderns must address these background issues before discerning any specific ethical principles. In other words, if we Moderns are to think through the Anthropocene, we will need a radical reordering of disciplines and boundaries within the university. And this re-ordering itself is, I argue, an ethical and political task. Similar to Eduardo Kohn's argument that the focus of anthropology must move beyond the human to the rest of the natural world,[4] so I would argue that the humanities must think beyond "the human" toward the rest of the natural world. Thinking the humanities ecologically, in other words, is the only way to re-ground truth and knowledge in the planet and revitalize the rest of the natural world. This chapter argues for at least three realignments within the university that will help retool the university and reimagine it as a technology that helps to create a planetary rather than an anthropocentric understanding of our world. Without rethinking the technologies of disciplined thinking patterns, we Moderns will be unable to create new habitats for acting, feeling, imagining, and valuing beyond the Anthropocene.

First, as I have argued elsewhere,[5] Moderns must cease to see "science" and "religion" as somehow separate entities that come into dialogue or conflict from time to time and understand them rather as always and already together. They are siblings who are always co-defining one other and cannot be separated out by some sort of understanding of a neutral, secular space. In this sense, I argue along with Latour, we have never been Modern.[6] Second, as James Miller argues in his recent book, Moderns must begin to decolonize epistemology.[7] This means paying attention to other than Western systems of knowledge production, decentering the sciences as the measure against which all knowledge must be judged, and paying attention to wisdom and perspectives from the more than human world. Third and finally, the university must become a place that reimagines the human being as first and foremost a planetary creature among other creatures. Thinking through the Anthropocene requires that we humans begin to think of ourselves, first and foremost, not as national citizens or as only a specific culture, race, and religion, but as planetary creatures. Before jumping to these more constructive conclusions, I begin here by analyzing some of the technologies that help to create and maintain the Anthropocene.

Technologies of the Anthropocene: the fossil-fueled era

When I use the word "technology" throughout this chapter, I mean it in the broader, Ancient Greek sense, which can mean language, ideas, art, and anything that helps us to shape the material world around us. Technology is, in this sense, part of being human. Technology is not limited to human persons, but human persons seem to rely on it to inhabit the world to a greater extent than any other persons or beings on the planet that we know of (for better and for worse). Technologies in this broad sense shape the becoming of Earth bodies in different ways (both good and bad). One technology found in monotheistic traditions, the idea that humans alone are created in god's image, has arguably been used to justify sexism, racism, and anthropocentrism and is in part responsible for the contemporary ecological crises.[8] Though there are many technologies that led to the development of what some call the Anthropocene (and which I would rather call the fossil-fueled era),[9] including the history of technologies that promote human exceptionalism from the rest of the natural world, I want to focus here on three within the history of "Western" thought: the objective and subjective divide that gives rise to a "smooth" secular space (critique of science and religion)[10]; the hyper-individualism that leads to solipsistic and colonizing epistemologies (critique of epistemology); and the hyper-capitalism that treats the rest of the natural world as a source of technology transfer for (certain) humans (critique of human exceptionalism). In doing so, I provide a contrast to the next section of this chapter on "New Technologies for the Planetary." In particular, I want to focus on the different political-ethical implications of these different technologies and highlight the radically different worlds they help co-create.

Objectivity, subjectivity, and secularity and the science and religion divide

As Foucault and so many other "post" thinkers have pointed out, the idea that truth (and true knowledge) is somehow in a space removed from power and politics (either in the reductive below or the transcendent above) is, in part, due to the burial of the process of the co-construction of ideas over time.[11] As such, truths and ideas that are "taken for granted" seem foundational: the processes of their co-construction have been hidden. Such seeming foundations become the ground from which we build our realities. In order to uncover these foundations and the power-filled ways these foundations get formulated, Foucault offers two methods for "doing" intellectual history: the genealogical approach and the archeological approach. The one (genealogy) seeks to uncover the contested origins of ideas while the other (archeology) seeks to look at the layers of competing ideas, discourses, and institutional structures that are co-formulated along with it.[12]

To offer but one example that I am more familiar with, we might examine the doctrine of *ex nihilo*.[13] The doctrine of *ex nihilo* is not a biblical doctrine. It can be traced to arguments between early church fathers and in particular those who were arguing against Gnosticism. Most Gnostic church fathers were looking at the world through bi-focal lenses: there was the material world and the spiritual world, which had corresponding evil gods and good gods, respectively. Furthermore, the Hebrew Testament was about the evil god and the Christian Testament was about the good god. Tertullian, among others, posited an "ex nihilo" creation in part as a way to say that there was one reality that included both the material and spiritual worlds, and that it was all good. This was codified in the Nicene Creed in the fourth century.[14]

However, the idea of *ex nihilo* creation was also taken up into colonial projects. It became the objective point (first cause/unmoved mover) by which all other effects and movements must be understood. It was the bulwark against infinite regress in epistemological claims. In other words, if there is a creator of everything seen and unseen out of nothing, then how could belief in any other metaphysical, epistemological, cultural, ethical, and moral ideas amount to anything other than wrong? Those holding these "other" (false) beliefs must, then, be "educated" into the one true Reality. Even when monotheistic revelation is replaced by objective truths reached through reason, and God is replaced with Nature (in the long, ongoing process called the scientific revolution), the basic model of the truth as one remains. In other words, the so-called scientific revolution and subsequent enlightenments replace God with Nature, revelation with reason, and import the idea that truth is one; the main difference is that science rather than religion (theology) becomes the preferred vehicle that takes us to that truth. Furthermore, enlightenment/scientific humanism imports the theological idea that humans are exceptional or uniquely distinct from the rest of the natural world, that nature is "below" humans, and that we are managers of the rest of the natural world. The transcendent nature of God, and the place of humans above the rest of the natural world, becomes justification for transforming passive nature through science.

In addition to this process of making nature the space of neutral objective science, religion becomes more and more about subjective, value-laden, private experiences. This is a sweeping statement, of course, and there are plenty of exceptions, but the idea that the religious life is part of the personal/subjective/ value-laden side of reality and science, economics, law, and politics belong to the objective/public/ neutral side of reality becomes the justification for spreading this "secular" idea through the globalization of neoliberal capitalism. The very split between personal/subjective and objective/public is something that is located in a subjective, Western modern mentality that is projected as objective. What is then, in actuality, something internalized historically from culturally located monotheism—the truth is one, humans are above the rest of the natural world, nature can be transformed by humans without much moral concern—becomes the basis for projecting and imposing these ideas on the rest of the planetary community as "objective" realities. The implications these formal structures have for ethics and politics are enormous and become the structures from which Moderns articulate anthropocentric ethics (for Moderns, simply Ethics). In addition to these subjectively located understandings and categorizations of reality, there is also a type of solipsistic individualism that is smuggled in from certain versions of theological anthropology.

Solipsistic individualism and colonial epistemology

One of the main sources of the hyper-individual in Western understandings of the human is found in the theological anthropology of the Omni-God. It turns out that the collective projection of an all-powerful, all-knowing, and ever present God may have more to do with the desires of people living amidst great uncertainty thousands of years ago than anything else. In other words, in a world where the life span was short and there were no modern technologies that "protected" us and (seemingly) sealed us off from the rest of the natural world, we needed to believe that some sort of all powerful deity was on our side protecting us. As Feuerbach and many others have suggested,[15] this theological projection is born of an anthropological desire: less rational, more affective. Through conceptually securing ourselves off from the rest of the natural world and through making our individual selves in the image of that Omni-God, we begin to shape the world around us as if we (some humans at least) are these little individual Omni-gods, and we call this process *reasoning*. In other words, at the boundaries of our reasoning are human desires cloaked in sub-rational foundational principles. The Cartesian and Lockean understandings of the human being are just further manifestations of this understanding of the individual.

Because the individual self is, whether or not admittedly so, embodied in the world in specific ways and because racism, sexism, classism, and ableism (among other markers of embodiment) provide unearned opportunities and adversities in a given time and social situation, certain individuals (usually white, male elites) have often become the ideal example of the hyper-individual. These ideal exemplars background their dependency upon human and Earth-others,[16] and because they accumulate greater wealth and power over time, they are able to more easily live "as if" they are these little Omni-gods. The technology of so-called "free market" capitalism (which is actually a welfare state for the wealthy) further enables the accumulation of wealth and power for these individuals and again, helps to set up a false projection that "good" or "proper" individuals are self-contained, singly responsible for their own lot in life, and are separate from the rest of the natural world, which is merely a resource for individual ends.

Hyper-capitalism and dead nature: human exceptionalism

If humans are thought to be separate from and exceptional to the rest of the natural world and certain hyper-individuals project and enforce their own subjective experience as the objective space of secularity and neutrality, then the "free market" mentality (as technology) is what enables the continuation of the system at the hidden expense of many poorer humans and the rest of the natural world. Many readers are likely familiar with the global "champagne glass" economy, where the world's top 1 percent control about 50 percent of the world's resources and wealth. As Val Plumwood has noted, it is usually also the case that the top 1 percent (or even 20 percent) are the ones who hold the most political power. These people are the most removed from their decisions ecologically and socially because their wealth shields them from the consequences of their actions.[17] Wealth enables one to "background" the labor and resources one depends upon for that wealth. It enables one to background the agency of multiple others that lead to a particular individual's accumulation of wealth: human, animal, biotic, abiotic, historic, systemic, and so on.

From the top of the champagne glass looking down, one does not see the stem or the base of the glass holding the whole thing up: one just seems to have gotten there or always been there on one's own talents, efforts, and value. One can be nearly blind to the histories of unearned privileges and discriminations that enable certain bodies to rise to the top, while keeping others on the bottom.[18] It becomes easier, then, to believe the narrative that "a rising tide lifts all boats," that one can "pull oneself up by the boot straps," or that "there is no alternative" (TINA) to hypercapitalism. This places the focus on individual responsibility and individual effort and erases alternative global economic systems, while ignoring the systemic agencies at work that heavily influence where one ends up in life. In a sense, it is the production of wealth by corporations and financial markets that serves as a kind of god who keeps the wealthy from falling down into the abyss. The more one lives by the market and succeeds, the more one becomes like a corporate self: seemingly all powerful, living longer and larger than any one individual entity is able to (eternal and omnipotent), and capable of moving about the globe freely regardless of local customs and constraints (omnipresent), and so on. This is a form of corporate Theo-anthropology, and it is at the root of the so-called American Dream mythology.

The problem with all of this is, of course, that no one is an island: all planetary entities are relational through and through. In fact, what we call individuals are more like ecosystems made up of multiple relationships past, present, and future, human and non.[19] We are porous bodies,[20] and for porous bodies, isolation and disconnection mean death. Indeed, the imagined separation by those on the "top" of the global economy (an abstraction from the ongoing flows of the planetary community) causes much death and destruction to Earth bodies. What we need, then, are technologies that help us to become more aware of our porous, planetary entanglement. We need to co-create different formal structures that will enable us to think about character, politics, aesthetics, value, and ethics from within a planetary perspective.[21]

New technologies for the planetary

The good news about destructive technologies is that they can be tinkered with over time and changed. Granted, technology is not something that humans merely use; we are transformed by technology and each technology changes how humans relate to other humans and to the rest of the natural world. Being born and raised in the US, I can no more imagine a world outside of the television and telephone, than those born in the US from about 2000 on can imagine a world outside of the Internet and cell phones. These are not just "things" we use, but they shape the ways in which we think and relate to ourselves, one another, and the rest of the natural world. We are, as Haraway argued decades ago, already cyborgs.[22] One of reasons to understand ourselves and our reality as cyborgian is to realize that the divisions between self-other, human-animal, plant-animal, biotic-abiotic, active-passive, and subject-object are categories imposed upon a much more complex reality. These mutually exclusive binarisms set up a "middle space" that is neutral, no-man's land, in-between. This is, I argue, the space of the secular. Unlike, for instance, the more porous thinking found in Daoism, where each side of the spectrum or distinction contains the other (yin-yang),[23] in Western, Modern thought, the law of non-contradiction has meant mutual exclusivity or a type of radical dualism.[24] These Modern categories uphold the secular divide and are enabled by the secular divide. Technology from this perspective, for instance, is value neutral and something that can be used by humans, or not, according to this divide. A gun, accordingly, is just a technology: it is the human who holds all agency in terms of how, when, and if it is used. Humans, unlike all other objects on the planet, hold agency. From within the secular space, science, technology, and economics are only fodder for human ends; they cannot tell us or shape what those ends ought to be. As argued earlier, however, the divisions enabled by the secular divide do indeed shape the world in value-laden ways that favor some life, ideas, bodies, and dreams over others.

In this section, I offer three different challenges that might help cross the divides reinforced by Modern technologies so that we might begin to think more in terms of a planetary community, which requires a planetary polis and a planetary politics and ethics. Whatever exists within the planetary cannot be outside of it and cannot be separated one from the other: value, objectivity, opinion, fact, subject, object, agency—all of these things are mixed up together within the planetary. Three catalysts for helping us to break through the Anthropocene (and all of its binary divisions) are a rethinking of "science" and "religion" as always mixed up together, thinking with multiple perspectives beyond colonizing Western epistemology, and breaking down the disciplinary boundaries in the university (one of the most important producers of the technologies that co-create worlds from generation to generation) in order to rethink ourselves within a planetary context.

Beyond the secular divide: science and religion as maps[25]

From within a planetary perspective, there are no neutral spaces. The idea of a "secular," neutral space depends upon the idea that truth and knowledge transcend politics, power struggles, histories, cultures, opinions, and values. In other words, it depends upon the old idea that the truth is somehow "out there" beyond the fray of day-to-day existence. From this perspective, right knowledge (becoming Enlightened, or perhaps better "en-whitened") will lead to right action. What happens when we understand ourselves as inextricably intertwined with the evolving, planetary community?

Ernst Haeckel, a 19th-century Romantic German scientist, understood some of the epistemological and aesthetic implications of placing humans within an evolutionary perspective. He argued (against Kant) that there could be no *a priori* ideas or values because humans evolved from the rest of the natural world: our knowledge emerges from the ongoing mix of planetary evolution; it is not written in the stars or contained in transcendent forms.[26] For this reason, he argued that there are no bare facts and that all knowledge involves hermeneutics and interpretation of the world around us.[27] For him, value and beauty were also emergent features of the world; thus we ought not look toward the realm of transcendent forms and ideals to find beauty and truth, but rather we should turn our eyes toward the many art forms emerging from nature.[28] It is the study of nature, not scripture or theology, that will lead us to collective notions of goodness, beauty, and truth.

Precisely because non-human nature is the reason for our very own existence, it is an active agent and it is value-laden, through and through. Scientific study, for Haeckel, must also eventually lead to discussions of philosophy, ethics, value, and even religion. Haeckel's religious naturalism, which he called Monism, was not separate from the sciences but was another aspect of the study of nature. Religion and science, for Haeckel, were not at all separate entities, split by some secular space that enabled "science" and "religion" to come together in dialogue; rather, they were different maps for understanding the natural world around us.[29]

Though we should certainly not agree with all of Haeckel's conclusions, the worst of which were based upon hitching religious naturalism to nationalism, we can learn from his placement of humans into an evolutionary perspective. As Baird Callicott notes, the problem with Romanticism was not its nature-based spirituality and philosophy, but that it was hitched to nationalisms or localisms. If we hitch a critical Romanticism to a planetary perspective, rather than to a nationalism, or localism, or to humanism alone, then everything on the planet is included in that vision.[30] If we are emergent from the rest of the planetary community, then whatever we call science and whatever we call religion are both features of that entire community—not just some within that community. They are not separate but multiple ways of mapping out planetary terrains.[31] They must also not be endeavors confined to humanity: other animals use tools; other animals have language and experience awe; perhaps even plants have communication or "feeling" of sorts.[32] What we experience as distinct methods for exploring the world are found to some degree throughout the planetary community, and we would not be human without the entire planetary community to think with. As James Miller states, "It takes a planet to make a human."[33] If this is the case, then we have a lot of work to do in terms of decentering and decolonizing epistemologies that are based upon the singularity and unity of truth and knowledge.

Decolonizing epistemology: multi-perspectivalism and non-absolutism

The idea that truth is one, that certainty is preferable to uncertainty, and that knowledge is progressive have perhaps done more harm than good.[34] Certainty in theological and religious ideas has helped to fuel crusades, colonization, wars, and hatred of "the other" in general. Certainty in scientific ideas—of what is "natural" or "true" scientifically—has also been used as a justification for enlightening the "dark masses"; for dictating what is healthy, sane, and normal; and for continuing the colonial project under the guise of progress and development. If truth is one, then certainty has few other options than proselytization. Those who are not informed of the one truth must be educated, assimilated, subjugated, or in the worst case eliminated. One has an ethical duty, from within this monological perspective to "spread the word."

A switch from a view of truth and knowledge as progressive and singular to a view of truth and knowledge as meandering/nomadic understanding and one that is based on multi-perspectivalism and non-absolutism (or toward a knowledge that depends upon uncertainty) is not a move to relativism, alternative facts, or "fake news." Haraway argues this point well when she writes that universalism and relativism are two sides of the same coin: they both offer knowledge regardless of context. Contextual, situated knowledge and a refractive (prismatic and multiple understandings rather than reflective singular) understanding of truth pay deep attention to our embodied, located, planetary existence.[35] Many feminist philosophers and critical theorists argue for this multi-perspectival understanding of truth. I won't go through that history here, but I will offer a couple of ways of thinking about it.

The first one comes from the Heisenberg vs. Bohr interpretation of quantum physics. As Karen Barad has argued: most scientists have followed Heisenberg, when perhaps we should be following Bohr.[36] Whereas Heisenberg thought that we just didn't know enough to understand how quanta could act like both wave and particle, depending upon the experiment, Bohr argued that reality is fundamentally indeterminate. For Heisenberg, eventually science will get to "the truth" that will explain earlier confusions: knowledge is progressive, and it progresses toward the true reality. For Bohr, in contrast, reality is fundamentally indeterminate, and it matters how we interpret it.

Barad extrapolates from Bohr's insights, using Foucault and Butler, to suggest something like even the quantum level of reality has some sort of agency, all of reality is co-created performatively, what we understand as natural "laws" are performances that have become habits or truth regimes shaping the worlds around us (not out of nothing but in a context), and different truth regimes are just that: different. She does not mean that anything goes. We do exist on "common grounds," which also depend upon the "undercommons" of those common grounds.[37] We humans are all on the same planet, we are animals, we breathe oxygen, we are subject to gravity, and so on. These are all grounds "for a time" that can shift and change; they are not eternal foundations, and we as part of the ongoing, evolving planetary community could never reach such certain foundations. But we can cobble together common grounds, common grounds that recognize the moral worth of all entities involved and that recognize the exclusions of those common grounds (the undercommons) on which those grounds appear. Such cobbling takes and benefits from multiple perspectives: it must pay attention to the buried "undercommons" that hold up the very ground on which we stand.[38] In other words, it must take into account the many subjugated voices and histories.

A second way of thinking about this type of contextual, planetary truth is found in an old parable from Vedic traditions. This story, as it is told from the Jain perspective, represents the concepts of non-absolutism (*syadvada*) and multi-perspectivalism (*anekantavada*). The paraphrase of the parable is something like the following:

> There are three blind men, each standing next to a different part of an elephant. One of them, standing by the elephant's leg, says: "An elephant is sturdy, and built like a tree trunk." The other, standing by the elephant's ear says: "An elephant is flat and thin." The third, standing by the tusk of the elephant says: "An elephant is long, hard and pointed."

The point, of course, is that we are all like one of those three blind men: we are part of a larger whole that we cannot see. We are always interpreting the information we are gathering from within our contexts. Each of us has only partial truth and partial knowledge, but together, these multiple perspectives provide a better overall map for a given territory. The production of knowledge and truth is a collective and ongoing or evolving process. It involves deep "listening" to human others and Earth others. We can learn a lot about the world from the perspectives of how other animals experience and sense the world. We can "learn" from other animals, plants, and entities what the "good," "true," and "beautiful" world might be, given what it needs for those entities to thrive. We can also learn from that which we don't know, recognizing that the holes, hidden things, and dark energies and matter are just as important as those things seen. Knowing should not be akin to discovery and enlightenment but rather to an ongoing project of co-creating worlds together. Decolonizing epistemology and opening on to a planetary understanding of knowledge (and aesthetics and ethics) means, finally, that we must also think about education and the humanities beyond the human.[39]

Education for the planetary: education beyond the human

My primary argument at the end of this chapter is to argue that education is through and through not only political but ethical. The character and values that are imparted through the educational system are affective and tacit and have much more efficacy and agency than arriving at formal principles through philosophical deliberation. History, psychology, what is healthy, what is insane, how we relate to the rest of the natural world and people with perspectives other than our own: these are all things that are implicitly if not explicitly influenced and organized through the ways in which the educational system is set up. In terms of developing a sense of ethical awareness, education is more about character and virtue development than it is about utilitarian calculations or universal principles. Before arguing for the need for what we might call "planetary humanities" in the educational system,[40] here, I take a quick detour to explain modified understandings of Aristotelian causality and reason that better meet the planetary context.

Aristotelian causality was not limited to efficient causality as much of Western, Modern knowledge is today. Aristotle's understanding of reason was not confined to instrumental reason, as much of Western knowledge is today. These larger understandings of causality, and of reason, are intertwined. For Aristotle, there were four causes: material, efficient, formal, and final. Roughly, material causality has to do with the shape and structure of a thing in itself; efficient causality is "A causes B"; formal causality has to do with the more meta-structure of things, or the ways things are shaped and put together; and final causality has to do with end goals.[41] In terms of looking at the educational system, Modern, Western education more and more focuses on the efficient cause to the detriment of other causes, for example, the purpose of education is to prepare you for a degree to get X job. I argue that we should think of education more as a formal cause than as an efficient cause. Education as a system shapes the lives and values of persons. As a formal cause, it might better be understood as embedded within a final cause of the flourishing of the planetary community (the final cause). Education is, then, not just about the human individual's economic well-being or even about the well-being of humanity alone, but it is for the flourishing of the planetary community. Instrumental reason is important here, but instrumental and practical reason are subordinate to wisdom, which has its focus on living within the overall planetary community. In this sense, the movement of thinking and causality does not go from individual outward but from the planetary community back to the individual. What would a humanities education from the perspective of planetarity look like?

Breaking Down Disciplines

In 19th-century Europe, the "naturalistic" worldview had not yet been fully formed, and many scholars were trying to figure out how to relate different strands of knowledge into some sort of coherent understanding of the world. The "culture wars" between emerging scientific disciplines (biology, geology, evolution, botany, and so on) and religious authority (theology) helped shape education into the primary disciplinary split we know today: humanities and sciences. At that time, however, both were still considered sciences (*Geisteswissenschaften* and *Naturwissenschaften*). Many tried to understand the human sciences and the natural sciences as a single tapestry that would eventually make up some sort of coherent view of the world. The "Renaissance" type thinkers of the era (Marx, Durkheim, James, Humboldt, Haeckel, and Goethe, among many others) spanned what we now identify as several different disciplines precisely because disciplinary boundaries had not yet been solidified. Unfortunately, this was also the time of the Industrial Revolution and science (*scire*, to take apart) eventually took on a more reductive and productive meaning in that context. Eventually, the reductive and productive model, with the help of quantitative social sciences, suggested that the human sciences could be reduced to the natural sciences, all of which could eventually be reduced to the physical sciences, and all for the purposes of developing better technologies for human beings in the realm of production, communication, transportation, and medicine. This reductive and productive model of nature results in what some sociologists and historians are calling the Great Acceleration since World War II: the period of huge anthropic impacts on the planet due to the increase in speed of communication, production, and transportation technologies.[42] I argue that such fossil-fueled and nuclear-fueled speed depends upon a "dead nature" that can be reduced to human ends ("standing reserve").[43] In other words, a reductive and productive model of nature and an understanding of nature as dead stuff for use toward human ends meets the need of the globalization of neoliberal capitalism quite well.

This reductive and productive model is good for producing things—policies, laws, medicine, energy and agricultural technologies, for instance—but it is not so good at answering more general questions about what types of worlds we want to live in and what the good life might be. The disciplines (even in the humanities) have become silos, to such an extent that the specific languages, theories, and tools of a given discipline are almost unrecognizable to someone outside that discipline (much less outside of the academy). Disciplines are then reduced to their own conversations, and at least within universities, multi-disciplinary thinking and teaching is not institutionally supported well. Students go to school to learn a trade and to get a specific job—not necessarily to learn and wonder about the world in general. Furthermore, the disciplines are geared mostly toward the human world and the needs, desires, hopes, and dreams of humanity. What gets funded and what does not often reflects the productive outcomes of a given research project for human beings (or at least some human beings).

In order to think planetarily, we need to break apart our disciplinary silos and start to think about planetary humanities, and science and technology for the planetary community (not just the human).[44] In a sense, we need to articulate what an education system might look like in a post-humanistic world: one in which the measure of truth, goodness, and beauty is hitched to the thriving of the planetary community and not just the human. This type of formal structure (planetary pedagogies and curricula) toward the end of the thriving of the planetary community will help to create the habitats for growing the ethical character and virtues needed for the health and well-being of the planet.

Breaking Down the Public/Academic Divide

It is not enough to deconstruct and reconstruct disciplines within the university, but we Moderns must also tackle the problem of the great divide between academics and broader publics. Though historically, this divide has been much greater, it takes on a whole new meaning in an era of instant communication. For most of human history in most cultures and places, education was geared toward learning familial trades and professions that were necessary to maintain daily life. Historically, only the very wealthy (or religious) were literate and/or learned in what we might now call academic subjects. This changed a great deal (at least in the Western world) with the Reformation in Europe, when more people were encouraged to read scriptures for themselves in their own languages. Literacy became important in order to read scriptures for oneself, but of course, that also meant reading other things. It changed even more as public education began to take hold. Even so, the knowledge imparted through education was mostly produced by an elite few. In the United States, at least, knowledge production began to shift dramatically with the GI bill, which allowed many returning soldiers after World War II to become the first generation in their families to attend college. In fact, the liberation movements of the 1960s and 1970s were (in part) fueled by university students who were beneficiaries (either directly or through their parents) of this opening up of the university to different races and classes, genders and sexes. The postmodern turn began to take hold in part because there were people of different genders/sexes/sexualities, of different races, and of different economic backgrounds challenging the dominant narratives within institutions of higher education (and within political and social institutions). As many universities became havens for critical inquiry and breaking down prejudices based upon race, sex, sexuality, and so on, certain populations within the United States (and elsewhere) began to turn away from these challenges, and the university was eventually painted as a place of liberal, left thinking.

This is a sweeping and broad overview, but it gives just a small nod toward the current situation of higher education politically, at least in the US (and in some European countries). Alternative facts and "fake news" are, ironically, supported by gross misinterpretations of postmodern thought, while the academy has become so insular and so involved in its critical theories and languages that it has not done the job of translation necessary to be relevant to those outside of the university.

Hence the anti-intellectualism on the populist right: the university is for elite, snowflake liberals, and the real working folks don't need such garbage; they just need to go back to some idealized "older" way of living and thinking, one which was good for (white male) people. At the same time, the anti-intellectualism on the left has taken hold because of the gap between the intellectual halls of the university and politics and favoring action over deliberation and thinking.

Any educational system geared toward planetary flourishing must also be overtly political. Education is always political, but deconstructing and reconstructing disciplines and exchanging knowledge between the larger public communities and the university is an ethico-political mandate in an era where a return to nationalism or neoliberal globalization as usual threatens both global political and ecological security. This blurring between the academic and public worlds will also allow for multiple ways of knowing to flourish.

Opening up to Multiple Ways of Knowing

Part of decolonizing epistemology[45] and re-thinking education for a planetary community is opening up to multiple ways of knowing. This means, on the one hand, that we ought to end the "scientification" of all knowledge.[46] As science is the favored measure of all knowledge in the contemporary global community— it gets the most funding and has the most political clout—most other forms of knowledge have been in the game of being more "scientific." Everything must be assessed and measured and have specific, efficient outcomes that lead to technology transfer and economic gain.[47] This scientific hegemony is challenged by many but especially two areas of inquiry: postmodern critiques/critical theories and now climate change denials (and some religiously motivated deniers of evolutionary theory). It is through the debates about evolutionary theory and, more important, climate change that science itself has to acknowledge its political process of knowledge production. Science marches on Washington and the persistent refrain "science is real"[48] reveal that science does not provide an objective view of the world that automatically compels one to the truth. Rather, it is also created and "done" in communities: with histories, sexualities, sexes, genders, politics, classes, and so on. This in no way means that "anything goes." Rather, as postmodern critiques argue (against both those who might deny climate change outright or those who blindly follow whatever science tells them), it simply means that a better way of understanding truth is to look at a thing, event, phenomenon, or situation from the point of view of multiple perspectives and disciplines (even from the perspectives of non-human life). As Bruno Latour argues, scientists might be better understood as representatives of the non-human world than dictators of that reality.[49] On the one hand, we can argue that physics, chemistry, biology, psychology, philosophy, history, and religion all provide different maps of the common planetary grounds we inhabit. On the other hand, we can say that the perspective of different peoples from different places of different genders, sexes, and races and the perspectives of other animals, plants, and life systems also offer us maps for the common planetary grounds we inhabit and help to co-create. These are common grounds that will shift and change over time, but they are stable enough to build worlds upon "for a time."[50] The ethical implications of those worlds should be monitored for the ways in which their constructions ripple out to effect different Earth bodies differently. We need cartographies of these effects in order to monitor the violence they cause and in order to re-think them to continuously help alleviate that violence. This is an ethico-political process that goes beyond the human, beyond any given generation, and beyond any final or single solutions. It is nothing less than a process of co-creating the ongoing evolving planetary community. The questions are what types of planetary communities do we want to help co-create, and how do we get there?

Notes

1 On "human exceptionalism," see: Anna Peterson, *Being Human: Ethics, Environment, and Our Place in the World* (Berkeley, CA: University of California Press, 2001).

2 I explore these connections further in: Whitney A. Bauman, *Theology, Creation, and Environmental Ethics: From Creatio ex Nihilo to Terra Nullius* (New York, NY: Routledge, 2009).

3 Andreas Malm and Alf Hornborg, "The Geology of Mankind? A Critique of the Anthropocene Narrative," *The Anthropocene Review* 1 (2014): 62–69.

4 Eduardo Kohn, *How Forests Think: Toward an Anthropology Beyond the Human* (Berkeley, CA: University of California Press, 2013).

5 Whitney Bauman, "Religion, Science and Globalization: Beyond Comparative Approaches," *Zygon: Journal of Religion and Science* 50, no. 2 (2015): 389–402.

6 Bruno Latour, *We Have Never Been Modern* (Cambridge, MA: Harvard University Press, 1993).

7 James Miller, *China's Green Religion: Daoism and the Quest for a Sustainable Future* (New York, NY: Columbia University Press, 2017).

8 For a good re-examination of the infamous Lynn White Hypothesis, see Todd LeVasseur and Anna Peterson, eds., *Religion and Ecological Crisis: The "Lynn White Thesis" at Fifty* (New York, NY: Routledge, 2017).

9 For critiques of the Anthropocene nomenclature, see, e.g., Malm and Hornberg, "The Geology of Mankind"; Lisa Sideris, *Consecrating Science: Wonder, Knowledge and the Natural World* (Berkeley, CA: University of California Press, 2017); and Axelle Karera, "Blackness and the Pitfalls of the Anthropocene Ethics," *Critical Philosophy of Race* 7, no.1 (2019): 32–56.

10 On the "smooth space" of secular objectivity, see Gilles Deleuze and Felix Guattari, *A Thousand Plateaus: Capitalism and Schizophrenia* (Minneapolis, MN: University of Minnesota Press, 1980).

11 Michel Foucault, *The History of Sexuality*, Vol. 1 (New York, NY: Random House, 1978).

12 Michel Foucault, *The Archeology of Knowledge and the Discourse on Language* (New York, NY: Tavistock, 1972).

13 Bauman, *Theology, Creation and Environmental Ethics*.

14 Ibid., chs. 1–2.

15 Ludwig Feuerbach, *The Essence of Christianity* (London, UK: John Chapman, 1854); David Noble argues this type of projection is also the basis of understanding technology religiously; see his *The Religion of Technology: The Divinity of Man and the Spirit of Invention* (New York, NY: Penguin, 1997).

16 On the colonial process of "backgrounding," see Val Plumwood, *Environmental Culture: The Ecological Crisis of Reason* (New York, NY: Routledge, 2002).

17 Ibid., 62–80.
18 Peggy McIntosh, *White Privilege: Unpacking the Invisible Knapsack* (Wellesley, MA: Wellesley College Center for Research on Women, 1988).
19 Connie Johnston argues this at both the microbial and climate levels of reality in her chapter, "Gut check," in this volume, "Climate Change Ethics and the Non-Human World." Amanda Nichols also argues for this "post-human" turn in the humanities in her chapter for this volume, "Being Human," as does Sam Mickey with his chapter on "Atmospheres of Object-Oriented Ontology."
20 Miller, *China's Green Religion*; Myra Rivera, *Poetics of the Flesh* (Durham, NC: Duke University Press, 2015).
21 Developing such a planetary identity is the topic of my book: *Religion and Ecology: Developing a Planetary Ethic* (New York, NY: Columbia University Press, 2014).
22 Donna Haraway, *Simians, Cyborgs and Women: The Reinvention of Nature* (New York, NY: Routledge, 1991).
23 Zairong Xiang, *Queer Ancient Ways: A Decolonial Exploration* (Earth, Milky Way: punctum books, 2018).
24 Val Plumwood, *Feminism and the Mastery of Nature* (New York, NY: Routledge, 1993), 41–68.
25 On epistemological maps, see, e.g., Mary Midgley, *Science and Poetry* (New York, NY: Routledge, 2001); and Peter Harrison, *Territories of Science and Religion* (Chicago, IL: University of Chicago Press, 2015).
26 Ernst Haeckel, *The Wonders of Life: A Popular Study of Biological Philosophy* (New York, NY: Harper and Brothers, 1905), 5–11.
27 See, e.g., the following claim: "No science of any kind whatever consists solely in the description of observed facts," (Haeckel, *Wonders of Life*, 5–6).
28 On this point, see Marsha Morton, "From Monera to Man: Ernst Haeckel, Darwinismus, and Nineteenth-Century German Art," in *The Art of Evolution: Darwin, Darwinism, and Visual Culture*, eds. Barbara Larson and Fae Brauer (Hanover, NH: Dartmouth College Press, 2009).
29 Ernst Haeckel, *Monism as Connecting Religion and Science: A Man of Science* (New York, NY: Dossier Press, 1919).
30 Baird Callicott, *Thinking Like a Planet: The Land Ethic and the Earth Ethic* (New York, NY: Oxford University Press, 2014).
31 Harrison, *Territories of Science and Religion*.
32 See, e.g., Marc Bekoff and Jane Goodall, *Minding Animals: Awareness, Emotions and Heart* (New York, NY: Oxford University Press, 2002); and Peter Wohlleben, *The Hidden Life of Trees: What They Feel, How They Communicate* (New York, NY: Harper Collins, 2017).
33 James has said this to me many times in conversation, and it is also one of his conclusions in *China's Green Religion*.
34 Catherine Keller, *God and Power: Counterapocalyptic Journeys* (Minneapolis, MN: Fortress, 2005).
35 Donna Haraway, "Situated Knowledges: The Science Question in Feminism and the Privilege of Partial Perspective," *Feminist Studies* 14, no. 3 (1988): 575–99.
36 Karen Barad, *Meeting the Universe Halfway: Quantum Physics and the Entanglement of Matter and Meaning* (Durham, NC: Duke University Press, 2007).
37 Laurel Kearns and Catherine Keller, *Ecospirit: Religions and Philosophies for the Earth* (New York, NY: Fordham University Press, 2007), 1–20. On the "undercommons" of those common grounds: Stefano Harney and Fred Moten, *The Undercommons: Fugitive Planning and Black Study* (Autonomedia, Online, 2013).
38 Catherine Keller, *Political Theology of the Earth: Our Planetary Emergency and the Struggle for a New Public* (New York, NY: Columbia University Press, 2018).
39 See, e.g., William Connolly, "Toward an Eco-Egalitarian University," *Class, Race and Corporate Power* 2, no. 2 (2014): 1–15.

40 James Miller and I are developing a "manifesto" of sorts on the planetary humanities.

41 Aristotle, *Metaphysics*, in *The Complete Works of Aristotle*, edited by Jonathan Barnes, vol. 2 (Princeton: Princeton University Press, 1984).

42 J.R. McNeill and Peter Engelke, *The Great Acceleration: An Environmental History of the Anthropocene Since 1945* (Cambridge, MA: Harvard University Press, 2016).

43 Carolyn Merchant, *The Death of Nature: Women, Ecology and the Scientific Revolution* (New York, NY: Harper and Row, 1980); Martin Heidegger, *The Question Concerning Technology and Other Essays* (New York, NY: Harper and Row, 1977).

44 Whitney Bauman, "Incarnating the Unknown: Planetary Technologies for a Planetary Community," *Religions* 8, no. 65 (2017): 1–10.

45 Ada Maria Isasi-Diaz and Eduardo Mendieta, *Decolonizing Epistemologies: Latina/o Theology and Philosophy* (New York, NY: Fordham University Press, 2011).

46 Kocku von Stuckrad, *The Scientification of Religion: An Historical Study of Discursive Change, 1800–2000* (Berlin, Germany: de Gruyter, 2014).

47 Rebecca Lave, Philip Mirowski, and Samuel Randalls, "Introduction: STS and Neoliberal Science," *Social Studies of Science* 40, no. 5 (2010): 659–75.

48 Lisa Stenmark, "Modern Political Lying: Science and Religion as Critical Discourse in a Post-Truth World," in *Navigating Post-Truth and Alternative Facts: Religion and Science as Political Theology*, ed. (Lanham, MD: Lexington, 2018), 3–18.

49 Bruno Latour, *Politics of Nature: How to Bring the Sciences into Democracy* (Cambridge, MA: Harvard University Press, 2004).

50 Thomas Tweed, *Crossing and Dwelling: A Theory of Religion* (Cambridge, MA: Harvard University Press, 2008).

8 Conflicting advice

Resolving conflicting moral recommendations in climate and environmental ethics

Patrik Baard

Climate ethics and environmental ethics

Two central problems currently facing the global community are climate change and biodiversity loss. The division of labor in applied ethics entails that the two challenges are discussed primarily in the separate fields of climate ethics and environmental ethics, though of course exceptions exist (Callicott 2013; Jamieson 2014; Rolston 2012).

This chapter will use a minimal definition of environmental ethics and climate ethics, despite both containing many different orientations. More specific strands of environmental ethics will be referred to when relevant in the chapter. Normatively, a minimal definition of environmental ethics is *ethical frameworks consistent with recognizing non-anthropocentrism*, that is, that not solely humans have moral standing. The minimal definition is consistent both with environmental ethics that identify the source of intrinsic values in ecosystems, or species (Rolston 2012), as well as those that consider humans as the source of value but where intrinsic values of ecosystems are the loci of values (Callicott 2013), in addition to frameworks that do not consider ecosystems to have intrinsic value but recognize their important instrumental weight to sentient beings (Attfield 2015). While climate ethics often utilize human-oriented normative concepts, such as rights (Humphreys 2010), or preference and welfare utilitarianism of climate economics, there is less explicit discussion on the applicability of such concepts to non-human beings. Climate ethics and even more so climate policies operate primarily from anthropocentric axioms or at least require us to withhold judgment on the moral worth of non-human organisms, as discussions of different forms of values of non-human organisms and states of affairs, which are at the core of environmental ethics, are often omitted in climate ethics and policies.[1] Thematically, environmental ethics investigates the moral standing of both individual animals and other organisms as well as ecosystems and species, or *nature* in a broad sense. Climate ethics do not necessarily have to consider such themes but may remain limited to human welfare.

The preceding definition does not imply that environmental themes are categorically rejected by climate ethics nor that climate issues are ignored by environmental ethics. But it specifies a division of labor. This division of labor would not be a problem were it not the case that the two policy areas affect each other. Ambitious climate change policies will have substantial impact on global species richness. Conversely, reduced biodiversity may result in more vulnerable eco-systems with reduced resilience to adapt to climatic changes. The adjacency and mutual influence of these policy areas and the separation of their ethical spheres are the main background of this chapter. I suggest that the two fields of applied ethics may propose conflicting action guidance, due both to differences in the focus of the issues analyzed and to different axiological premises.

The lack of environmental ethical concern in climate ethics and policy may be understandable, as climate change, in a sense, *trumps* biodiversity loss in terms of the scale of harms that will result. Biodiversity loss, even if it constitutes a moral wrong, will not result in as great a risk to all forms of life as the dangers posed by climate change. Yet, climate policies that recognize environmental ethics are likely to have more overall value than those based exclusively on anthropocen-trism, as environmental ethics entails reverence for the lives of both future genera-tions and non-human organisms.

One should not underestimate the alarming and urgent nature of climate change. However, that urgency does not permit one to implement *any* measure (see, for instance, Preston 2016 for discussions on the ethical aspects of geoengineering proposals, such as solar radiation management). It is imaginable that a climate ethics uninformed by environmental ethics may go to great extents to permit interventions as long as they are believed to avoid or mitigate dangerous climate change. This includes permitting the destruction of habitats and ecosystems rich in biodiversity; such destruction may sometimes be justified if required for reduc-ing dangerous climate change, but it should be ethically informed.

Climate change policy and biodiversity loss: incompatible conclusions and conflicting advice

This section first presents an overview of moral conflicts and disagreements understood as conflicting recommendations from two viewpoints of well-considered moral judgments. It then discusses, throughout Section 2.2, the conflict at hand in this chapter.

A note on moral conflicts

Not every moral framework allows for moral conflicts; utilitarianism, for instance, does not. If there are two options that will result in equal amounts of utility, there is no moral conflict; rather, one is permitted to remain agnostic about which to choose. However, virtue ethics and, to a lesser extent, deontic ethics commonly permit moral conflicts, even if deontic ethics will in some instances suggest that moral conflicts are often only apparent, since just one moral demand may ulti-mately be considered the proper moral demand.

Moral conflicts concern instances when (at least) two moral demands apply, but both cannot be fulfilled, forcing a moral agent to make a choice to forfeit fulfilling one demand. Usually, the moral agent will not be exempt from the for-feited demand, despite its no longer being possible to fulfill (Williams 1973). Thus, when breaking a promise to meet a friend in order to help someone who is injured, the promise to meet the friend is a prima facie demand but may result in a residual obligation, such as an apology or an explanation. In moral *tragedies*, such constructive expressions of restitution or repair may not be available, and all that remains is to express grief over an irreversible loss.

Moral disagreements, however, concern the validity of moral judgments. In instances of moral disagreement, how are we to settle the disagreement? How do we determine that (at least) one of the judgments is invalid or wrong, and what viewpoint would grant us confidence in such an assertion? We can surmise that *if* a moral judgment stems from or is consistent with a well-considered judg-ment or with other judgments that one adheres to by living accordingly, we can at least have some, even if inconclusive, confidence in that moral judgment and its reasonableness.

Avoiding moral conflicts (or moral disagreements) is not always possible. Indeed, many of our demands stem from interpersonal relations, and avoiding con-flicting demands would entail avoiding having any relation with another human (or, indeed, any other morally relevant being, including non-human organisms and possibly ecosystems) at all (Hansson 1998). Yet, a moral conflict as such does not stifle action. Rather, it calls for a more ethically conscious action, for knowing what one is forfeiting and understanding the reasons why. When a wrong has been committed, as when one neglects to fulfill a demand, one may have a duty to repair that wrong, even if the wrong occurred due to one's fulfillment of another duty.

For example, if a habitat must be transformed into a monoculture for carbon capture, that may be permissible if it fulfills the demand of reducing climate change risks, even if the transformation transgresses several environmental ethi-cal demands, such as by reducing biodiversity.

Climate change and biodiversity: conflicts and convergences

Currently, climate change policies that are the least dangerous, understood as not leading to global mean temperature increases above 1.5–2 °C relative to the 1850–1900 time period, require both reducing the sources of greenhouse gases—emissions—and extensive transformation of habitats into croplands to use for bioenergy, along with transforming land for negative emission technologies like forestation, direct air capture technologies, and ocean fertilization. That is, it will require both reducing fossil fuel emissions and sucking carbon out of the atmosphere. The least risky option that the IPCC has presented, RPC2.6,[2] will likely result in reduced global species richness due to these land use changes (Newbold et al. 2015). In their report on the prospects of limiting global warming to 1.5°C by 2100 relative to pre-industrial times, the IPCC write that the large transitions required to maintain within 1.5°C "pose profound challenges for sus-tainable management of the various demands on land for human settlements," including biodiversity (IPCC 2018, 18).

But are there policies that can satisfy both climate and environmental ethical demands, to limit the risks of dangerous climate change while at the same time preserving biodiversity? The choices that are investigated here as being potentially consistent with both climate and environmental ethics are

1   Reducing emissions, supplemented with substantial use of negative emission technologies by transforming habitats.
2   Limiting human population growth.
3   Technology efficiency and development.
4   Limiting consumption of energy through policies and behavioral changes.

I argue that only the fourth choice is consistent with both environmental ethics and climate ethics (see Section 2.2.4).

There are three variables to keep in mind in the following analysis: population size, the use of energy and land per capita, and total consumption. If population growth is reduced, consumption of energy per capita may remain at current levels and thus entail reduced consumption of total energy and land requirements relative to a scenario in which population continues to grow. However, population growth would be consistent with decreased total consumption of energy and land requirements if the consumption of energy per capita is reduced, as by developing efficient technologies. Energy is primarily, but not solely, understood as entailing the use of fossil fuels, whereas land requirements will be affected by both energy uses like bioenergy and other consumption uses such as agriculture and animal husbandry. I will treat habitats and atmospheric capacity to hold carbon as commons to which all need access but where excessive use will result in dangerous climate change and accelerate the sixth mass extinction. It should be noted that substantial decreases in the total consumption of energy and land use are required to avoid dangerous climate change but that the model proposed seeks to abstract the different factors at play. While this is admittedly a great simplification of the mechanisms underlying anthropogenic climate change and unprecedented biodiversity loss, one aspect that justifies the simplification is that current consumption levels, a consequence of both population size and consumption of energy, and land requirements put enormous pressures on both the atmosphere through fossil fuel emissions and on habitats through increased land transformations.

I start with an overview of the current choice to reduce the risks of dangerous climate change, as analyzed by the IPCC, to provide a sense of why it is problematic from an environmental ethical perspective.

Current trajectories entail conflicting moral judgments

The IPCC has developed different RCPs for their analyses. In their fifth assessment report, the RCP with the least risk for dangerous climate change was RCP2.6, which has a likely mean temperature increase range for 2100 of 0.3–1.7 °C relative to 1850–1900 (IPCC 2013). It is the only scenario that is not likely to exceed 2 °C increases and is thus the safest option. RCP2.6 will not only require reductions in emissions but also net negative emissions. In other words, limiting sources of greenhouse gases (emissions) is not enough; it must be supplemented by other strategies such as "negative emission technologies" (Fuss et al. 2016),[3] including but not limited to bioenergy with carbon capture and storage, afforestation and reforestation, direct air capture and storage, soil carbon sequestration, and ocean fertilization (Fuss et al. 2016). However, such technologies have significant uncertainties in terms of the extent of negative emissions that will be realized and unintended socio-economic and environmental consequences of their large-scale deployment (Fuss et al. 2016).

The recent investigation by the IPCC on the prospects for limiting global mean temperature increases to 1.5 °C by 2100 relative to pre-industrial times evokes similar issues (IPCC 2018). Without going into too much detail, reaching a 1.5 °C target requires increased pressures on the sectors *agriculture, forestry, and other land use* (AFOLU) (IPCC 2018). AFOLU includes policy areas such as food and feed production, wood production, producing biomass, and Carbon Dioxide Removal (CDR) being a process of deliberative removal of carbon dioxide from the atmosphere and placing it in reservoirs (IPCC 2018, 144). The pathways to 1.5 °C all balance these different needs (IPCC 2018, 16). For example, one pathway envisages lower energy demand up to 2050, through rapid decarbonization, where afforestation is the only required CDR option. In contrast, a resource- and energy-intensive pathway in which there is a high demand for transportation fuels and livestock products entail that emission reductions are mainly achieved through strong use of CDR and deployment of carbon capture and storage that can affect biodiversity (IPCC 2018, 16). In the latter pathway to 1.5 °C, extensive efforts are required for carbon removal. In the former pathway, less intervention is required beyond afforestation, due to rapid decarbonization and lower energy demand.

One essential part of RCP2.6, and of 1.5 °C, is the increased usage of land for biomass and reforestation and negative emission technologies. Furthermore, novel approaches such as algal biofuels are being investigated. In recent years, technologies such as artificial trees have also been developed; these are manufactured trees with increased capacity to store carbon from the air, a process known as direct air capture (see Preston 2018 for a discussion on some ethical aspects of artificial trees). The second least risky RCP, RCP4.5, has an expected temperature increase range of 1.1–2.6 °C (IPCC 2013). That scenario thus entails greater risks of dangerous climate change but will nonetheless require environmental interventions similar to those utilized in RCP2.6, albeit on a less substantial scale.

It is estimated that the landmass required for forestation to have any impact on levels of atmospheric greenhouse gases would be equivalent to three times the size of India (*Guardian* 2017). It is unlikely that such efforts would have negligible environmental impacts. For one thing, it risks creating large-scale monocultures for biosequestration and thus reducing biodiversity because of preferences for tree species that can rapidly absorb carbon from the air. While some estimates suggest that mixing tree species would absorb carbon just as well (Hulvey et al. 2013), even a tree mixture that avoids monocultures may still lead to biodiversity losses. Of course, it makes a difference whether such plantations would be on agricultural sites or in rainforests. In the latter case, there would be a substantial biodiversity loss, even if tree mixes are in a sense better than monocultures. However, given the substantial need for land areas for afforestation to have any effect on the concentration levels of greenhouse gases, forestation will most likely affect both former agricultural sites and rainforests.

Novelties such as artificial trees, as well as ambitious large-scale forest plantation projects, surely evoke intuitions of what has natural value and what our obligations to the environment are. The natural environmental is increasingly becoming intentionally designed to fit human purposes, and we may ask whether something of value is lost. Indeed, negative emissions technologies have been described as "anthropogenic activities that deliberately extract CO_2 from the atmosphere" (Fuss et al. 2016). Yet, it is not only that environmental areas are designed but also that artifacts such as artificial trees are installed in the forests, with entire ecosystems transformed for the sake of reducing the risks of dangerous climate change (Preston 2018). Even if we grant that something of value is lost, such as reduced wilderness, biodiversity, or other natural value, we have to estimate whether that loss is permissible if required to reduce the harms of dangerous climate change. Given enough risks of morally relevant losses if such policies are not implemented, they may even be obligatory.

We find a central conflict in RCP2.6 and the ambition of limiting global mean warming to 1.5 °C. Since they rely on a substantial conversion of habitats into croplands for biofuels and enabling carbon sinks, they would entail a reduction in global species richness. Indeed, according to some estimates, RCP2.6 would lead to a greater reduction in global species richness than all other RCPs, with the exception of RCP8.5, which is the riskiest of all and has been called "business as usual" in the sense that it does not foresee any climate-relevant policies being implemented (Newbold et al. 2015). However, RCP2.6, as well as 1.5 °C, requires "rapid conversion of primary vegetation, especially in the tropics, to crops and biofuels" (Newbold et al. 2015, 48). In contrast, RCP6.0 and RCP4.5 entail less biodiversity reductions than the other alternatives but greater risk for dangerous climate change.

Yet, to willingly, intentionally, and knowingly support policies that will contribute to biodiversity loss and likely species extinction is inconsistent with conservation and respect for nature, being central demands to some fields of environmental ethics. If we assume that species loss as such entails a great harm, then the favored trajectories are problematic since they rely so heavily on changing land use with a negative effect for global species richness. To reach this conclusion requires that one consider species, or ecosystems, as such to have moral standing (see Rolston 2012 for an example), alternatively that respect for nature is virtuous (Jamieson 2014).

Given the ongoing sixth mass extinction, with species going extinct at a rate a hundred or even a thousand times naturally occurring extinction rates (Wilson 2016), it is questionable whether it would be permissible or in any way reasonable to choose an alternative that would further accelerate extinction rates. It may indeed be permissible, given that climate change will also entail biodiversity loss; however, since RCP2.6 is the only option that will not risk breaching 2 °C increases in mean temperatures by 2100 but will lead to the greatest loss of global species richness of the RCPs (with the exception of RCP8.5), we seem to be caught in a moral tragedy with two conflicting demands, where forfeiting either one will result in irreversible losses. We can either choose ambitious climate policies and by doing so accelerate the sixth mass extinction, or we can implement ambitious policies to halt or slow that extinction, at the cost of increasing risks of dangerous climate change, since halting the sixth mass extinction would require not implementing ambitious climate policies. Climate ethics is consistent with substantial reductions in biodiversity if required to reduce the risks of dangerous climate change, as long as additional factors are also warranted, such as ecosystems resilient enough to provide anthropocentric utilities. But that climate ethical recommendation would be inconsistent with environmental ethics, since it does not respect or give appropriate weight to the interests of non-human organisms and states of affairs.

This disagreement takes the form of supporting two opposing judgments, which can be phrased as "it is obligatory that biodiversity be preserved" from an environmental ethical position and "it is not obligatory that biodiversity be preserved" from a climate ethical viewpoint. In a more measured version, the disagreement involves choosing between priorities of biodiversity preservation and ambitious climate change policies; these could be phrased as "it is not permissible to contribute to biodiversity loss even if required for ambitious climate change policies" from an environmental ethical perspective and "it is permissible to contribute to biodiversity loss if required for ambitious climate change policies" from a climate ethical perspective. The propositions can be extended to include fuller arguments and suggestions, but the core disagreement regarding what normative judgment that should apply to the environment and, consequently what forms of reverence for nature are required, will persist.

It is suggested in the following that we should develop an environmental-ethically concerned climate ethics (see for instance Jamieson 2014; Rolston 2012). Although the sheer scale of climate change means that we will most likely have to make substantial sacrifices, it would then be virtuous to at least express regret over those sacrifices, as well as being virtuous to try to minimize the losses. First, however, two other options are investigated.

Limiting population growth

The central problems discussed in this chapter are the substantial pressures on the atmosphere and on habitats due to human activities. Intuitively, one way to reduce those pressures would be to reduce population growth, since population growth can safely be assumed to increase such pressures. Thus, both climate ethics and environmental ethics are prima facie consistent with reducing population growth.

The approach of reducing population growth to reduce pressures on limited resources has been favored by, for instance, the Club of Rome (Meadows et al. 1972), as well as biologists like Paul Ehrlich (1978). Overpopulation has also been discussed in relation to environmentalism (Cafaro and Crist 2012). Limiting population growth would theoretically have the benefit of maintaining current per capita consumption levels while reducing pressure on atmospheric capacity and habitats, since overconsumption and overpopulation are two motors in the extinguishing of animal and plant species (Gambrel and Cafaro 2010).

According to some authors, overpopulation is due to anthropocentrism, and "people's apparent willingness to live in an ecologically devastated world" (Crist 2012, 149). On another assumption, population growth is needed to ensure productivity and standard of living (Simon 1981). But even so, there are several reasons to be wary of reducing population growth, even if one assumes that it would mitigate emissions and reduce pressures on land use. One reason is normative: there may be a justified intuition that there is something morally troublesome about such policies related to, for instance, non-identity problems and the repugnant conclusion (Parfit 1984) and would thus be problematic from anthropocentric premises. The non-identity problem is, together with the repugnant conclusion, a central challenge to population ethics, from which we can detect arguments for population growth (Attfield 2015, 125).

The problems were both formulated by Derek Parfit (1984) and concern the moral standing of future, currently non-existing, individuals and populations. Parfit imagines a current choice between two policies regarding some resource, *Depletion* or *Conservation*, where the former will entail that the quality of life for any chosen member of future generations is much lower than in the latter but still worth living. If we choose the policy *Conservation* now, those future persons that exist if we choose *Depletion* will not exist. Therefore, it is difficult to make the claim that we have hurt the future groups of persons existing in *Depletion* (Parfit 1984, 361ff). Had we not chosen *Depletion*, they would not have existed at all.

The repugnant conclusion is summarized by Parfit in the following manner:

> For any possible population . . . all with a very high quality of life, there must be some much larger imaginable population whose existence, if other things are equal, would be better, even though its members have lives that are barely worth living.
>
> (Parfit 1984: 388)

The reason for the latter being better is due to it containing an accumulated greater utility and is thus consistent with utilitarian ethics of maximizing utility, despite the individuals having lives barely worth living. Rights-based theories face similar problems regarding future, but currently non-existent, individuals (Attfield 2015: 37). Parfit does not want to accept this conclusion, hence labelling it repugnant.

Relevant to the topic of this chapter, it would be questionable whether anthropocentrism would permit reducing population growth for non-anthropocentric reasons and even more so since it does not even permit reducing population growth for solely anthropocentric reasons. Moreover, the two problems solely concern persons and human quality of life, granting solely instrumental value to other entities.

Assumedly, a strong environmental ethics could be consistent with limiting population growth, if preservation of species and ecosystem stability have sufficient moral significance relative to anthropocentric interests. Perhaps the most notable example is deep ecology, which from Arne Naess's viewpoint regarded population reduction as a very long-range issue that had no prospect of being realized now (see Naess 1989: 140ff). But there are further reasons to reject this proposal besides being normatively troublesome. There is no agent, either global or multilateral, that is able to set such a policy in motion. Furthermore, the thesis lacks empirical support. Since technological development results in efficiency, developments in emission technologies mean that the same amount of energy can be produced with fewer emissions than previously and that the same amount of crops for food (or indeed biomass) can be produced using less landmass. Thus, let us turn to technological efficiency to see whether it may help with a solution that satisfies both environmental and climate ethics.

Technological efficiency

Much of our economy relies on fossil fuel emissions and the transformation of habitats to suit productive purposes. Recent years have witnessed an upsurge in "green technology" and renewable energy. If successful, these advances would result in maintaining current per capita consumption levels with a reduced impact, since they would rely on renewables like sun or wind power or on technology that produces the same amount of energy but with fewer emissions. These technologies are not without some environmental impact: wind power may affect ecosystems and habitats in the vicinity of wind turbines, while solar power may have a negative effect because of the minerals and hazardous materials used in producing solar panels and the land use required for solar power facilities. Intuitively, it is not difficult to imagine that further development of green technologies may limit such negative impacts. But does such an unqualified intuition have empirical support?

Theoretically, successful implementation of green technology strategies would decouple economic growth from increases in fossil fuel-based emissions and greater pressures on land use. At least this is the assumption that "absolute decoupling" is premised on. This would satisfy both environmental and climate ethical criteria, for it would lead to reduced pressure on habitats and the atmosphere while protecting economic prosperity. One could, ideally, have the same level of welfare and economic prosperity without increasing emissions and reducing habitats. One problem is that decoupling is often used as a relative concept. As Tim Jackson (2011) notes, the utilization of a new technology may seem green since it results in fewer emissions or lower habitat demands than a previous generation of the same technology, but it says nothing about absolute levels of emissions or ecosystem deterioration. That is, the technology as such may be more efficient but also be more widely used, resulting in greater emissions and reductions of habitats in absolute terms. Decoupling in absolute terms is, according to Jackson, not an empirical possibility (2011).

This is sometimes referred to as Jevon's paradox and entails that "improvement in energy efficiency causes exponential use of energy" (Higgins 2015, 130). The paradox strengthens the conclusion that the absolute decoupling of economic progress from emissions increases and habitat destruction is impossible. With the paradox in mind, relying on technological efficiency is unlikely to be consistent with either climate ethics or environmental ethics, since it would lead to increased pressures on both the atmosphere and habitats through increased emissions and transformation of habitats to productive uses, despite relative efficiency per produced unit.

Is limiting consumption consistent with both climate and environmental ethics?

It is suggested here that the pathway that is consistent with both environmental ethics and climate ethics and that will reduce pressures on ecosystems and the atmosphere is increased humility and a modified relation to consumption goods.

One common thread among the preceding options is that they intend to preserve high levels of consumption. Indeed, the discussed alternatives intend to keep per capita consumption of both energy and land use at current high levels but reduce overall pressure by reducing population growth (section 2.2.2) or to rely on technology efficiency (section 2.2.3). However, one additional alternative would be to substantially lower per capita energy use and consumption levels in order to preserve the atmosphere and mitigate land use pressures. That is, to reduce overall emission and consumption levels by reducing per capita levels. This would rely on an understanding of desires as insatiable beyond subsistence needs, a recognition that we rely on resources and a relation to our surroundings based on modesty.

Numerous philosophers have argued for modesty to express environmental reverence, which may prove instructive here. Henry David Thoreau argued for "simplifying needs and avoiding new wants" (see Purdy 2015). Similarly, John Stuart Mill admitted to seeing no pleasure in "contemplating the world with nothing left to the spontaneous activity of nature" and felt that there was great spiritual and moral progress to be made when "minds ceased to be engrossed by the art of getting on" (1965: 3:756). Mill regarded the best state for human nature as one in which, while "no one is poor, no one desires to be richer" (Mill 1965, 3:754). Both Thoreau and Mill insisted that the choice is always one where people could devote resources to put more pressures on land use and thus enrich themselves but choose not do so and conserve or preserve instead.

Both Thoreau's and Mill's suggestions are best understood as virtues, as forms of moderating one's wants and desires, or more accurately, making traits such as moderation and humility part of one's character and stable disposition. They are not proposing an ascetic ideal but rather modesty and humility, as expressed by Mill. Similarly, Thoreau suggested that wants will never be fully satisfied because

> he believed that no sum of satisfied appetites could bring lasting satisfaction, or heal pervasive anxiety, before people became clear on what they valued and why. These questions required reflection on who people were and how they fit into the world.
>
> (Purdy 2015, 141)

Virtue ethics asks us to consider the kind of people we should be (Sandler 2007, 27) and those qualities that make a person a good person (Gambrel and Cafaro 2010, 86). Admittedly, virtues are anthropocentric in the sense that they involve human characteristics, but virtues can certainly accommodate a concern and respect for non-human organisms and ecosystems (Sandler 2007). Several environmental ethicists have discussed virtues related to environmental ethics generally (Sandler 2007) and specifically related to climate change (Jamieson 2014; Gambrel and Cafaro 2010). Thus, Jamieson suggests that *green virtues* are called for in the Anthropocene (2014). Such virtues include humility, temperance, mindfulness, and cooperativeness (2014, 186ff). Especially relevant in the current discussion is the virtue *respect for nature*, for its otherness but also for instrumental reasons. To Jamieson, the virtues result in a general policy of reducing our contribution to global environmental problems.

In contrast, Gambrel and Cafaro base virtues of simplicity on a naturalistic account (Gambrel and Cafaro 2010, 87). Roughly, this means that virtues enable human flourishing and that "virtues are character traits that a person needs to flourish or live well" (Sandler 2007: 14). Traditionally, naturalistic accounts of virtues have claimed that flourishing is related to what species or life form an entity belongs to and to assess what type of being that entity is. In this sense, humans are rational social animals (Gambrel and Cafaro 2010, 87). From this follows that human virtues, the conditions for flourishing, include autonomy, meaningfulness, and good, functioning human communities, in addition to individual survival and other traits (Gambrel and Cafaro 2010, 88; Sandler 2007, 28). The virtue simplicity overlaps with temperance, frugality, prudence, and self-control; it requires active deliberation on consumer choices and needs to be cultivated by individuals but also encouraged and mandated by society and further justice by reducing overconsumption and unfair levels of consuming both goods and food (Gambrel and Cafaro 2010). However, simplicity does not entail a romantic "return to nature" or involuntary poverty.

A voluntary reduction of wants arguably offers one way of reducing the substantial pressures on both atmospheric capacity and land use, since, if implemented on a grand enough scale, they would lead to lower per capita and total consumption rates. Yet, reduced consumption is only one horn of the problem, since the vast inequity in access to and consumption of the utilities required for subsistence, such as energy and food, may persist despite those who have access to such goods choosing to limit their use of them.

However, to voluntarily reduce consumption could provide guidance for more modest lifestyles that would reduce emissions and ecosystem pressures, and a virtue such as simplicity would entail that more resources are left for others. One could understand Mill's suggestion of "no one desires to be richer" as proposing such simplicity. Moderation, together with simplicity understood as forfeiting to other morally relevant beings the goods that are not required for one's own subsistence but are for their subsistence, is consistent with both climate and environmental ethics, as it entails a reduction in consumption together with an increased concern and respect for our surroundings.

It is crucial to note that the virtues should not be confused with a "hands-off" approach that involves only forgoing actions that may affect nature. Many ambitious, large-scale efforts will be required to reduce the risks of dangerous climate change while also slowing the sixth mass extinction. As suggested earlier, achieving the negative net emissions required by RCP2.6 will require both large-scale afforestation and the implementation of technological novelties in addition to simplicity. The problem, however, is undertaking such actions in an ethically non-reflective manner. Such reasoning would benefit from viewing natural entities, such as non-human organisms and endangered species, as having more than solely instrumental value and as worthy of respect. That is, that their flourishing also has moral standing, and sacrificing their flourishing should not be unjustified or merely instrumental to our survival.

Virtue ethics could serve as a foundation for a general normative framework by which to formulate and implement policies. A framework of environmental ethically informed climate ethics, where economic growth and consumption levels are not regarded as general laws or overriding goals, requires looking toward other standards or values of moral progress, as Mill suggested. In such a framework, sacrifices such as the loss of ecosystems and of natural value will still be permitted if they are required to reduce risks of dangerous climate change, but the framework would not justify running amok thoughtlessly and without justification. Rather, it is a framework in which respect for nature plays a central role as a virtue, in which agents are aware of the sacrifices they are permitting, and in which they can enumerate reasons that justify those sacrifices. This would be an integral part of an environmental-ethically aware climate ethics.

Summary note

Four options have been discussed here. The first was transforming ecosystems and habitats to use for negative emission technologies, which would be consistent with climate ethics but not with environmental ethics. Limiting population growth would be consistent with environmental ethics but would be less compatible with the highly anthropocentric axioms of climate ethics. Technological efficiency would most likely increase pressures on both habitats and the atmosphere and thus be inconsistent with both climate and environmental ethics. Lastly, limiting consumption would be consistent with both climate and environmental ethics.

What I have investigated and cautiously proposed here is that virtues are consistent with reduced consumption levels of both energy and non-subsistence goods. These virtues can be supported by both climate ethics and environmental ethics, as they enhance respect for our current surroundings and for posterity. Although it could be argued that focusing on virtues is anthropocentric, it is better viewed as a precautionary approach to act *as if* non-human organisms and states of affairs are worthy of reverence and respect. Furthermore, it affirms that climate change trumps biodiversity loss in several ways, since climate change itself will lead to increased biodiversity losses. Therefore, the framework is consistent with implementations such as forestation, biofuels, and carbon capture, but such implementation has to be thoughtful and ethically justified. By including environmental ethical awareness, the proposed framework intends to supplement climate ethics by generating more environmental-ethically conscious decisions.

A virtue such as simplicity is consistent with saving both the atmosphere and resilient ecosystems for morally relevant entities, such as future generations and non-human organisms and states of affairs. The latter requires a climate ethics that is informed by environmental ethics. To express such virtues is consistent with reducing consumption levels of both energy and consumption goods and thus with satisfying both climate ethics and environmental ethics.

Notes

1 Climate policy sometimes includes biodiversity loss. Thus, for instance, the IPCC includes impacts on biological diversity in some of their assessments (IPCC 2014). However, to consider biodiversity does not necessarily qualify as recognizing non-anthropocentrism, as biodiversity assessments may view biodiversity as an instrumental good, like providing ecosystem services.
2 RCP, or *Representative Concentration Pathway*, is used to construct scenarios of future pathways by the IPCC (2018). The purpose of the RCP is to provide a framework by which to analyze long-term climate change and impacts of those changes. RCP include "time series of emissions and concentrations of the full suite of greenhouse gases and aerosols and chemically active gases, as well as land use/land cover" (IPCC 2018, 555).
3 In this chapter, "mitigation" refers to reducing emissions, that is, the source of greenhouse gases. The IPCC also includes the enhancement of carbon sinks as mitigation (IPCC 2014, 125). However, in order to focus on the conflict that is the topic of this chapter, it is more appropriate to separate the two, since enhancing carbon sinks entails transforming land use to enable negative emission technologies.

References

Attfield, P. *The Ethics of the Global Environment*. 2nd ed. Edinburgh: Edinburgh University Press, 2015.
Cafaro, P. and E. Crist, eds. *Life on the Brink: Environmentalists ConfrontOverpopulation*. Athens, GA: Georgia University Press, 2012.
Callicott, J.B. *Thinking Like a Planet: The Land Ethic and the Earth Ethic*. Oxford: Oxford University Press, 2013.

Crist, E. "Abundant Earth and Population." In *Life on the Brink: Environmentalists Confront Overpopulation*, edited by P. Cafaro and E. Crist, 141–59. Athens, GA: Georgia University Press, 2012.

Ehrlich, P. *The Population Bomb*. New York: Ballantine Books, 1978.

Fuss, S., C.D. Jones, F. Kraxner, G.P. Peters, P. Smith, M. Tavoni, D.P. van Vurren, J.G. Canadell, R.B. Jackson, and J. Milne. "Research Priorities for Negative Emissions." *Environmental Research Letters*, 11, no. 11 (2016). doi:10.1088/17489326/11/11/115007.

Gambrel, J. and P. Cafaro. "The Virtue of Simplicity." *Journal of Agricultural and Environmental Ethics* 23 (2010): 85–108. doi:10.1007/s10806-009-9187-0.

The Guardian. "Negative Emissions Tech: Can More Trees, Carbon Capture or Biochar Solve Our CO2 Problem?" 2017. Accessed July 27, 2018. www.theguardian.com/sustainable-business/2017/may/05/negative-emissions-tech-can-more-trees-carbon-capture-or-biochar-solve-our-co2-problem.

Hansson, S.O. "Should We Avoid Moral Dilemmas?" *The Journal of Value Inquiry* 32, no. 3 (1998): 407–16.

Higgins, P. *Eradicating Ecocide*. 2nd ed. London: Shepheard-Walwyn, 2015.

Hulvey, K.B., R.J. Hobbs, R.J. Standish, D.B. Lindenmayer, L. Lach, and M.P. Perring. "Benefits of Tree Mixes in Carbon Plantings." *Nature Climate Change* 3, no. 10 (2013): 869–74.

Humphreys, S., ed. *Human Rights and Climate Change*. Cambridge: Cambridge University Press, 2010.

IPCC. "Annex II: Glossary." In *Climate Change 2014: Synthesis Report. Contribution of Working Groups I, II and III to the Fifth Assessment Report of the Intergovernmental Panel on Climate Change*, 117–30. Geneva: IPCC, 2014.

IPCC. *Global Warming of 1.5°C*. Geneva: IPCC, 2018.

IPCC. "Summary for Policymakers." In *Climate Change 2013: The Physical Science Basis. Contribution of Working Group I to the Fifth Assessment Report of the Intergovernmental Panel on Climate Change*, edited by T.F. Stocker et al., 3–32. Cambridge: Cambridge University Press, 2013.

Jackson, T. *Prosperity Without Growth*. London: Earthscan, 2011.

Jamieson, D. *Reason in a Dark Time: Why the Struggle Against Climate Change Failed*. Oxford: Oxford University Press, 2014.

Meadows, D.H. et al. *The Limits to Growth*. Washington, DC: Potomac Associates, 1972.

Mill, J.S. *The Collected Works of John Stuart Mill, Volume III—The Principles of Political Economy with Some of Their Applications to Social Philosophy*. Books III–V and Appendices. Edited by John M. Robson. Toronto: University of Toronto Press, 1965. Accessed July 27, 2018. http://oll.libertyfund.org/title/243.

Naess, A. *Ecology, Community, and Lifestyle*. Cambridge: Cambridge University Press, 1989.

Newbold, T., L.N. Hudson, S.L. Hill, S. Contu, I. Lysenko, R.A. Senior, L. Börger, D.J. Bennett, A. Choimes, B. Collen, and J. Day. "Global Effects of Land Use on Local Terrestrial Biodiversity." *Nature* 520, no. 7545 (2015): 45–69.

Parfit, D. *Reasons and Persons*. Oxford: Oxford University Press, 1984.

Preston, C., ed. *Climate Justice and Geoengineering: Ethics and Policy in the Atmospheric Anthropocene*. London: Rowman & Littlefield, 2016.

Preston, C. *The Synthetic Age: Outdesigning Evolution, Resurrecting Species, and Reengineering Our World*. Cambridge, MA: MIT Press, 2018.

Purdy, J. *After Nature: A Politics for the Anthropocene*. Harvard: Harvard University Press, 2015.

Rolston, H. *A New Environmental Ethics: The Next Millennium for Life on Earth*. London: Routledge, 2012.

Sandler, R.L. *Character and Environment*. New York, NY: Columbia University Press, 2007.

Simon, J. *The Ultimate Resource*. Princeton, NJ: Princeton University Press, 1981.

Williams, B. "Ethical Consistency." In *Problems of the Self*. Cambridge, UK: Cambridge University Press, 1973.

Wilson, E.O. *Half-Earth: Our Planet's Fight for Life*. New York, NY: Liveright Publishing Company, 2016.

9 An eco-centric proposal for setting a price on greenhouse gas emissions

Karen Green

Aldo Leopold's advocacy for the introduction of a land ethic is now more than 50 years old, and since that time, others have added their voices to calls for a new environmental ethic.[1] Yet, despite theoretical support by environmental philosophers for the idea that we have obligations to nature and to land, those who have turned their attention to the question of the just allocation of rights to emit carbondioxide and other "greenhouse" gasses, have largely converged on the principle of a per-capita allocation of rights, with various concessions to historic responsibility and capacity to pay.[2] The idea that justice in the allocation to countries of caps on emissions is to be understood through equal per-capita shares has also become enshrined in policy documents. It is endorsed by Ross Garnaut in his report to the Australian government and is captured in the idea of "contraction and convergence."[3] International negotiations, so far, have assumed that the appropriate principle of justice in this case is one which begins from humans, and governments have been eager to insist on the rights of their citizens to economic development. Not surprisingly perhaps, the land ethic is nowhere to be seen.

Theories of justice have generally been anthropocentric, but the land ethic proposes that we have direct duties to land. One could see it as rediscovering an ethical relationship towards land that was, and remains, characteristic of the philosophical stance of the aboriginal peoples of Australia. From such a point of view, principles of justice among peoples will be constrained by duties towards the land on which people depend. So in what follows, I introduce constraints imposed by duties towards land into the question of just allocation of rights to emit greenhouse gasses and explore the prospects of an eco-centric response to the question of rights to emit carbon, based on geography. Since the compromises of the Paris Accord, there is every reason to be pessimistic about the prospects for a global emissions trading scheme. Nevertheless, in this chapter, I frame the question of a just allocation of rights to emit carbon dioxide and other greenhouse gases on the assumption that an international agreement on a global cap and trade system could be developed. This may well not be the case, yet it is becoming an urgent necessity to re-engage with the possibility of a binding international framework, and it turns out that the results of applying eco-centric principles for assigning rights to emit carbon are quite intuitive. The implications of taking the land ethic seriously involve developing limits to human rights to exploit an atmosphere on which all life depends and regimes which impose those limits.

The problem

The greenhouse effect is a classic version of the tragedy of the commons, which is made worse by its intergenerational features.[4] Each country and individual has a reason for using energy, and often this will create greenhouse emissions. The taking up and release of carbon (in the form of carbon dioxide) is part of the natural ecological cycle, but as the proportion of carbon dioxide and methane in the atmosphere increases, so does the globe's temperature, and feedback mechanisms tend to enhance this.[5] For many centuries, the number of people on the Earth and the emissions they produced have not affected the global atmosphere. But at some point in the 20th century, the unlocking of carbon through coal mining and oil extraction began to result in more carbon dioxide going into the atmosphere than could be absorbed back by the land and sea. The net annual increase of carbon dioxide in the atmosphere is dramatically represented in the Keeling curve.[6] It now appears incontrovertible that overexploitation of the atmospheric commons is leading to global warming and to the melting of the Arctic and Antarctic ice-sheets, an eventuality which will itself increase the rate of global warming. The worst-case scenario is runaway warming, which will result in the "Venus Syndrome," the evaporation of all water from Earth.[7]

In his classic paper on the tragedy of the commons, Garret Hardin argued that such tragedies can be solved by one of three methods: appeals to conscience, privatization, and systems of mutual coercion, mutually agreed on.[8] Without endorsing Hardin's unduly pessimistic views, in relation to the historical management of commons, which actually suggest that ethical principles of land use can be mutually agreed on and upheld by appeals to conscience, it also has to be recognized that appeals to conscience that are not enforceable are in danger of being ineffectual. In the case of the atmosphere, free-riding is just too tempting. In this case also, full privatization is impossible. This implies the need for mutually agreed on principles of use, backed up by sanctions for misuse. Since the atmosphere is global, there is no way the resource can be divided up, but some features of privatization can be achieved through tradable emission rights, which make countries pay others for the cost of emitting more than their "fair" share. While such a scheme is sometimes contrasted with a straight tax on carbon, in effect, it works as a more sophisticated form of global taxation, which should result in income transfers from high-emitting wealthy nations to low-emitting poorer ones and in which there is a market mechanism for determining the level of the tax. In order for an emissions trading scheme to work, the allocation of rights needs to be internationally agreed on and the rules for emissions trading to be settled under international law. This makes it imperative to consider the question of what counts as a fair share when we are considering rights to emit.

So far, the international community has adopted the strategy of attempting to develop some agreed on actions through diplomatic negotiation, but the resultant compromises have not resulted in strategies that can be relied on to solve the problem. Initially, the Kyoto Protocol agreed to the rather ad-hoc proposal that developed countries, called "Annex 1," should attempt to keep emissions to 5 percent less than their 1990 emissions. This was not scientifically justified. It was far too little to reverse the trend towards warming, and this feature of the protocol was exacerbated by the fact that no targets were set for developing countries. The strategy has implicitly embodied two conflicting principles: a need for cuts in emissions and the right of non-developed counties to pursue industrial development. Since the signing of the Kyoto Protocol in 1997, developed countries have made some cuts to emissions, but non-Annex 1 countries have made none. Instead, they have substantially increased their emissions. In acknowledgment of the fact that it is the countries that industrialized earliest that have put two-thirds of the carbon dioxide into the atmosphere, less developed countries only agreed to the protocol on the basis that their emissions would not need to be reduced for the time being. The outcome from Kyoto was that the US, which was responsible for between 18 and 22 percent of gross emissions, failed to ratify the protocol, arguing that it went against its interests. As a result of the Paris Agreement of 2015, developing countries agreed to begin stabilizing or reducing emissions after 2020, but the interim increases meant that by 2008, China, India, and the rest of the less-developed world (that is countries excluding Europe, Russia, Japan, Australia, and North America) already made up 50 percent of emissions, a growing proportion.[9] Current assessments of what is required in order to keep global warming to 1.5 °C, as published in the *IPPC Special Report*, October 2018, estimate that the globe needs to become carbon neutral, that is, to achieve zero net emissions by 2030. This report also explores the possibility of becoming carbon neutral by 2050, which would lead to overshooting the 1.5 °C target but would be compatible with achieving it by 2100.[10] The commitments made by signatories to the Paris Agreement fall far short of achieving even the less demanding of these goals. Indeed, the current US administration has withdrawn from the agreement, and less developed counties continue to attempt to negotiate non-binding national targets. So the need to articulate principles that could provide the basis for an alternative binding international agreement, acceptable to all, is urgent.

A number of philosophers have looked at the question of what principle ought to be adopted in order to allocate emission rights globally. One answer to the question of what would be a fair and effective emissions control regime follows the initially, intuitively fair principle that a fair share is an equal share.[11] Such egalitarian principles usually need to be modified to take into account differences in need and differences in desert, but they remain a good starting point for thinking about fairness. In what follows, I will first look critically at some of the arguments that have been offered for adopting the view that simple equal per-capita shares is the appropriate principle of justice for allocating rights to emit greenhouse gasses. I will then argue, on the basis of the existence of duties towards land and of considerations to do with differences in need and desert, arising from various kinds of historical responsibility, for an allocation based on land size. I conclude by showing that the consequences of adopting the principles developed are quite workable and consider some objections that might be raised.

There are at least three things necessary in order for the emissions control regime to be effective: 1) The total quantity of emissions needs to be set at a level such that the carbon cycle is stable—that is, the long-term trend of net emissions is zero. Or, more precisely, since this will not be immediately achievable, a time-table for the reduction of emissions to sustainable levels needs to be set. 2) The control regime must be practical in the sense that it is one that the parties will agree to. 3) It should include every country. If the regime is not universal, countries that are not part of the regime will simply undo the good work of the countries cutting emissions; therefore, only a global control regime can be effective.

Against equal per-capita shares

There are a number of reasons for thinking that a per-capita allocation can't fulfill these three desiderata. The first is that it involves perverse incentives with regard to population, and so, it does not set a fixed cap for each country. There is a sense in which, in the end, all situations that have the character of a tragedy of the commons are grounded in population growth. In each case, an activity that would be harmless were a small number of individuals involved becomes harmful because too many others are behaving in the same way. While the population of high carbon dioxide emitters was low, the Earth's systems could cope. It is growth in numbers of people achieving an industrialized, high-emission lifestyle that is the problem. So a solution should reward those countries whose carbon emissions go down because of declining population, as well as those whose emissions per head of population goes down. But a per-capita allocation may provide a perverse incentive to increase population, since, if it is difficult to reduce one's total out-put, one could reduce the per-capita consumption by increasing population. In order to avoid this problem, both Peter Singer and Dale Jamieson, who advocate per-capita allocations, propose setting a country's emissions per-capita relative to some date.[12] They see 1990 as a suitable date because it coincides with a period when the international challenge of anthropocentric climate change was clearly established. But allocating rights to emit on the basis of historical population numbers leads to a perverse incentive not to grow one's population, even if the situation requires it. Australia, for instance, might well be called on to take climate refugees, but if it were to be given a per-capita allocation that was fixed at 1990, it would not want to disadvantage its current population by further reducing their per-capita allocation. For this reason, Ross Garnaut accepts a per-capita allocation that simply follows population.[13] Yet this returns us to the problem that it builds in pressure to increase population. Further, it in fact penalizes counties that have adopted environmentally sound measures of population reduction.

For instance, in 1995, China had a population of approximately 1.24 billion, which is projected to be 1.4 billion in 2020 and is then expected to stabilize at this level. India, by contrast, had a population of approximately 900 million in 1995 but is projected to have just under 1.4 billion by 2020 and to continue to grow beyond that. In this situation, a per capita right to emissions in effect deprives China of some of the benefits that it expected to reap from its quite draconian population policies by making its allocation shrink as India's grows. This would be perverse, for it would fail to take into account one form of historic responsibility (responsibility for population growth), despite the fact that there has been agreement that a different kind of historic responsibility (responsibility for past emissions) has been enshrined in the principle that past polluters should pay. Underdeveloped countries have argued that since they did not cause the problem and have not benefited from past emissions, it is not fair that they should have to make cuts, since they will thereby be locked into underdevelopment. But this is not a reason for placing no constraints on underdeveloped countries. To do this is just to invite the transfer of emitting industries from developed to underdeveloped countries. The regime imposed should be universal but should also treat different countries differently, according to their different circumstances. One relevant difference is responsibility for one of the underlying problems—population growth.

It is right that those countries that have made the problem worse by emitting high quantities of carbon dioxide per head of population should do more than similar countries with lower emissions per head, but it is also right that those countries that have allowed their populations to grow should have lower per-capita emissions than similar countries that have taken steps to encourage smaller populations. From the point of view of the planet's capacity to absorb carbon dioxide, one million people emitting 25 tons per person is no different from 25 million people emitting one ton per person. This may seem like a quibble, but a country's population is a number that reflects certain historical choices. To take another example, in 1950, India had a population of just over 370 million, while the countries that make up the Russian Federation had a population of about 103 million. By 2010, India had grown to 1.2 billion, while there were just over 140 million in the Russian Federation. Should India, which once had less than four times the population of Russia, be given nearly ten times its allocation? It has been clear since the 1970s that there are ecological limits to the globe's capacity to support a growing population at high levels of consumption. Since the problem we face results both from high per-capita consumption and from high levels of population, both kinds of historical responsibility ought to be acknowledged.

There are arguments in the literature that purport to show that equal per-capita shares is the default fair option. Baer, for instance, uses the following analogy. Suppose that two people have both been exploiting a resource—he imagines an aquifer—but one has had the technology to take more than the other. The second person only acquires the technology to exploit the resource to the same level when it is starting to run out. Baer argues that the default fair allocation in this situation would be to share equally.[14] But the problem with this argument is that it is based on a defective analogy. The situation we find ourselves in is more like the following. One population has established a town on the side of a lake and has been using it to dispose of its wastes. Another population has been unable to access the lake and so has not been able to reap the advantages of cheap waste disposal. The first population has just begun to notice algal blooms in the lake and to realize that they are in danger of triggering its ecological collapse. They are facing a complete restructuring of their waste disposal in order to save the lake and their own civilization. But at this point, the second population acquires the means to dispose of their waste in the lake. They claim a right to the advantages the first population have benefitted from. But surely there are no rights to pollute. The first population needs to change its behavior. The second population does not have a right to make cleaning up the mess impossible by insisting on an equal liberty to pollute. The fact that others have *unwittingly* benefitted from an action that turns out to cause harm when too many act that way is not a justification for *wittingly* pursuing acts that cause this harm. This analogy suggests that, while it would be fair for the first population to shoulder all the expense of changing their habits and cleaning up the lake, the other population does not have a prima facie right to make the task more difficult by using it to dispose of their wastes. While there are arguably rights to a share of resources, there are no rights to pollute, only duties not to pollute. The difficulty is, of course, that, given current technology, it is difficult to exploit resources without polluting. But even if one looks at the problem as one of distribution of resources, it is not clear that equal shares is the appropriate concept of fairness to apply. We have already explored the way in which differences in historical population growth can imply differences in desert, which need to be balanced against the differences in desert that flow from historically higher per-capita emissions. There are other reasons also for thinking that equal shares is not necessarily the best criterion of justice to adopt, particularly in a complex economy.

In some way, it is odd that philosophers in the developed world have hit upon equal per-capita shares as the appropriate concept of fair distribution, in this instance. For the liberal capitalist market that they rely on for their economic well-being is instead committed to some version of what Rawls calls the "difference principle." This principle proposes that economic inequalities can be fair so long as they are to the advantage of the worst off.[15] Quite how large such inequalities should be allowed to grow and how large the benefits to the worst off need to be for the system to remain fair are complex questions at the heart of current political and economic debates, yet even the once communist China has been convinced that some inequality, private capital accumulation, and an entrepreneurial class who can deliver technological development are not unfair, if allowing them can deliver a higher standard of living to the poorest. So we can argue that, if some inequality in per-capita allocations of emissions rights is in the interests of the worst off, it will still be just. In fact, we will all be worst off if we do not achieve a global consensus on an equitable and workable distribution of the burden of reducing greenhouse gas emissions. So, if the price of a practical solution to the allocation of emissions rights is some inequality in the per-capita allocation, we should accept this. This argument might be challenged, in that it does not use the economic concept of "worst off" intended by Rawls. However, as we will see, given that the worst off are in general the very poor in Sub-Saharan Africa, a geographic allocation will turn out to be in the interests of the economically worst-off.

The last considerable problem with per-capita shares is that, given current global economic and politic realities, it is just not practical. The idea that the first world will contract its emissions so that they will converge towards those of the less developed world proposes that, by the year 2050, the per-capita greenhouse emissions of the developed world will equal those of the developing world. But this was simply not a reasonable demand. To achieve it would require the developed world to accept a 90 percent cut in its emissions in order to leave "room" for the developing world to ultimately cut its emissions by between 25 and 50 percent. It was little wonder that the US was not prepared to sign up to a treaty that required it to commit not simply to cleaning up its own mess but to making it pay for the mess that developing countries are continuing to exacerbate. The Rio Declaration, on which the Kyoto Protocol was based, proposed that countries should have common but differential responsibilities for addressing climate change, and this is no doubt broadly fair. But there are different ways of interpreting this vague phrase. Developing countries pointed to the fact that in 1990, North America, with only 5 percent of the world's population was responsible for more than 25 percent of the world's greenhouse emissions. This was surely more than its fair share. But was it practical to insist that it aimed to reduce its consumption in proportion to its population? In 2005, Asia had about 60 percent of the world's population, Africa 14 percent, Europe 11 percent, South America 9 percent, and Oceania a mere 0.5 percent. Per capita shares would require that North America cut its proportion of emissions to 5 percent, while Asia was allowed to grow from the approximately 30 percent it was then emitting to 60 percent. It is simply not clear how the US and Canada could afford to do this. It places a burden of cost on them to which the American people are simply not likely to agree, and it would in all probability be economically self-defeating, since the global economy is so tied in with the US economy. So it is worth at least considering whether an allocation based on land is feasible.

In favor of emissions allocations based on land

So far, philosophers who have dealt with the issue of allocating emissions rights have seen it as a problem of distributive justice involving humans. They have failed to think about the core problem, which is a problem of the over-exploitation of the environment. They have failed to sufficiently acknowledge that peoples have duties to care for the land that they occupy. In recent decades, some governments have begun to recognize that they have a duty not simply to the human population they represent but also to the land and environment that supports that population. We have ministers for the environment. Increasingly, the freedom of humans to pursue their economic self-interest has been limited by the requirement that the environmental integrity of a region be supported, a rare species protected, a wilderness preserved. Each government therefore can be seen as the contemporary representative of the land of the people who have elected it, as well as the representative of those people. It has the responsibility for maintaining its own population within the ecological limits of its own geography. If it does not do so, it is potentially infringing the rights of its neighbors and it is jeopardizing the long-term survival of the society that it temporarily represents.

Allocating rights to emit greenhouse gasses on the basis of geography simply extends such principles to the international realm. For simplicity, I shall assume that, on average, each reasonably large landmass has approximately the same capacity to absorb carbon dioxide. A more developed version of the proposal would no doubt modify this assumption. Those places that we call deserts are relatively impoverished in their capacity to trap carbon dioxide through photosynthesis, so a fair geographic assignment might have to take into account the nature of the land occupied. But most countries are made up of mixtures of fertile and more desert regions, so for the sake of argument, I will assume that countries made up of differing land types have on average the same capacity to act as a carbon sink. If we think that governments have a duty to their land, as well as to their population, they will have a duty to maintain the activities of their population within the capacity of their land to absorb carbon dioxide and other greenhouse gasses. Looked at this way, the fair allocation of rights to emit should be proportional to the country's land size and its inherent capacity to absorb greenhouse gasses. The land does not care whether this is achieved through a large population, which emits a small quantity per capita or a small population which emits a large quantity per capita. The important issue is that the country does not infringe on other countries' capacity to operate within their ecological limits, by producing more emissions than its land can absorb. In fact, every country can emit more than its landmass can absorb, because the sea is such a large carbon sink. But the seas are a genuine common. Apart from narrow coastal areas, they should not be privatized. Hence, in so far as a government is responsible for the environmental integrity of its own land and for maintaining its people within the ecological constraints imposed by its own geography, it ought not to emit more than is commensurate with its own capacity to absorb.

A geographic principle has the advantage that it builds in compensation for countries that choose to lock up potentially productive land as wilderness. The capacity of this uncultivated land to sequester carbon can then be sold to countries that need, at least in the short term, to emit more than their land can absorb. This gives countries a real choice as to whether to increase population, at the cost of decreasing the level of greenhouse emissions per capita, or to decrease population in order to find environmental space for increased per-capita consumption. It reflects the essence of the environmental problem, the fact that we have exceeded the capacity of the lands that we inhabit to absorb the greenhouse gasses that we are emitting.

The proposal turns out to be surprisingly practical, for the goal of bringing the proportional emission of greenhouse gasses in line with the geographic capacity for carbon sequestration should be easier to achieve politically than the goal of achieving equal per-capita emissions. In the early 2000s, North America, with 18 percent of the globe's landmass, was producing 29 percent of the fixed industrial emissions of carbon dioxide, while Europe, with 7 percent of the landmass and Asia with 33 percent were both producing a similar proportion. Africa, with

21 percent of the landmass, was producing only 5.4 percent as was Latin America with 13 percent. For North America to set the goal of reducing its carbon emissions to 18 percent of the global total is far more realistic than to expect it to reduce to 5 percent. Indeed, according to *Wikipedia*, by 2014, the proportion of greenhouse emissions produced by the US had reduced to about 15 percent, Europe's to 9 percent, while China's had increased to 26 percent and India's to 6 percent. These changes in proportion reflect the growth in emissions by the previously less developed regions, rather than any decline in the total. This direction of development has been counterproductive, since it should be easier for countries that are not yet dependent on carbon-intensive modes of producing power to develop using new technologies than for developed countries to completely redevelop their infrastructure. The less developed nations of Africa and Latin America need reliable sources of income to enable them to develop in a sustainable way. They, at the moment the worst off, will be better served under a geographic allocation than they would be under a per-capita allocation, because, relative to landmass, they are emitting so little. In fact, the geographic proposal may be somewhat harder on Europe than a per-capita proposal, but this is perfectly in line with Europe's historical responsibility. The area that is most affected by the proposal is Asia, which, on a per-capita regime would have a right to 60 percent of the world's emissions but which on a geographic allocation would only be entitled to 33 percent. But this too seems to appropriately build in responsibility for historic high levels of population growth.

The *Stern Review of the Economics of Climate Change* calculated that in order for carbon dioxide to stabilize in the atmosphere at 450 ppm by 2050, developed countries would have to have reduced their emissions by between 60 and 90 percent of 1990 levels, and developing countries also would have to have reduced their emissions by between 25 and 70 percent of 1990 levels.[16] These are going to be very difficult reductions to achieve, and scientists such as James Hansen subsequently argued that 350 ppm rather than 450 ppm is what we should be aiming for. So there is very little environmental room for increased emissions from developing countries. In a sense, one can take the proposed geographical allocations to be demonstrating this. As a proportion of the globe, Asia is already emitting its fair share and too much in absolute terms. In order to get in line with their fair shares, Europe and North America will have to substantially reduce their total emissions but are already close to emitting a fair proportion. Only Africa and Latin America have some room for growth, but it would be in everybody's interest if they were to invest the money they receive from selling carbon credits in educating their children, bringing down infant mortality, preserving their forests, and developing sources of renewable energy rather than perversely developing a carbon-intensive way of life that would immediately have to be wound back.

The proposal is practical in another sense. It obviates complex calculations and concessions for carbon sinks. Every landmass is a carbon sink, so this is already built into the allocation. The allocation of emission rights can simply be an allocation of gross emissions. The formula for allocating rights is simple. Once a time line for the contraction of the cap is set, countries will know that they will be getting a fixed proportion of the total, and a time table for the reduction of the cap can be set.

Some objections

I shall consider two objections: first, that the proposal will lead to some anomalous results, and second, the *ad hominem* objection that, as an Australian, I can't be serious in proposing this as a just allocation, since it is so blatantly in Australia's self-interest. It is true that the proposal implies some anomalies. There are a few very small densely populated countries and a few large very sparsely populated countries. The first would be out and out losers on this proposal. The second may seem to be unfair winners, particularly in those cases where they are desert countries, which by their nature are not very good carbon sinks. This might make one think that something like arable land ought to be the criterion on which to base the allocation, and no doubt a fully worked out proposal would have to take into account details of a country's geography. At the same time, given a shifting climate, it may be more practical to accept a relatively simple system of allocation. In the end, if Chad or the Spanish Sahara were able to sell carbon credits and use the capital to install solar power facilities that could provide clean power for Europe, so much the better. Those few places, such as Singapore, which are very densely populated, have to rely on other countries to produce their food and manufactured goods, so they should pay for the carbon emissions entailed by the goods they purchase from others. It is also true that, if one looks at individual countries, there may be significant problems with moving immediately to a distribution based on land, so it might initially be more feasible to adopt regional allocations, with country allocations to be determined within the region. Just as moving towards an equal per-capita situation would take a very long time, so too working towards the acceptance that densely populated countries should pay sparsely populated countries to preserve undeveloped areas to serve as carbon sinks will require significant political negotiation and change in political perspectives.

The *ad hominem* objection is a worry but might also be turned into a virtue. It is not simply because of Australia's interests that I think this proposal should be considered. The aboriginal peoples of Australia have in the past seen themselves as belonging to the land. Duties to care for country are central to their world view. Constraining anthropocentric rights by duties to land is then a local tradition and one that I sincerely think should be globalized, since it is the globe's incapacity to absorb the combined pressure of human populations that is the fundamental problem. Given global trade, the products produced with emissions expended in one country may be consumed a long way off, so that rather than complaining about a few anomalous cases, such as Australia, apparently getting off lightly, the global community ought at least to consider that historical responsibility for population growth has to be taken into account, alongside historical responsibility for high carbon emissions. In the light of continuing denial and the ineffectual remedies that continue to emerge from international negotiations, one can become pessimistic about the prospect of the international community acting seriously on climate change. The thoughts aired here at least show that it is possible to apply the principles of a land ethic to this practical ethical and political issue with surprisingly workable results.

Notes

1 Aldo Leopold, *A Sand Country Almanac* (Oxford: Oxford University Press, 1966).
2 See for instance Peter Singer, *One World* (Melbourne: Text, 2002), 44–57; Paul Baer et al., "Greenhouse Development Rights: A Proposal for a Fair Climate Treaty," *Ethics Place and Environment* 12 (2009), 267–81; Paul Baer, "Equity, Greenhouse Gas Emissions, and Global Common Resources," in *Climate Change Policy: A Survey*, eds. Stephen H. Schneider, Armin Rosencranz, and John O. Niles (Washington, DC: Island Press, 2002); Dale Jamieson, "Climate Change and Global Environmental Justice," in *Changing the Atmosphere: Expert Knowledge and Environmental Governance*, eds. Clark Miller and Paul Edwards (Cambridge, MA: MIT, 2001); Carsten Helm, "Distributive Justice in International Environmental Policy: Axiomatic Foundation and Exemplary Formulation," *Environmental Values* 10 (2001): 5–18.
3 Ross Garnaut, *The Climate Change Review. Final Report* (Cambridge: Cambridge University Press, 2008).
4 Stephen M. Gardiner, "A Perfect Moral Storm: Climate Change, Intergenerational Ethics and the Problem of Moral Corruption," *Environmental Values* 15 (2006): 397–413; "Ethics and Global Climate Change," *Ethics: An International Journal of Social, Political, and Legal Philosophy* 114 (2004): 555–600; "The Global Warming Tragedy and the Dangerous Illusion of the Kyoto Protocol," *Ethics and International Affairs* 18 (2004): 23–39; "The Real Tragedy of the Commons," *Philosophy and Public Affairs* 30 (2001): 387–416.
5 James Hansen, *Storms of My Grandchildren* (New York: Bloomsbury, 2009).
6 Ibid., 116.
7 Ibid., 223–36.
8 Garret Hardin, "The Tragedy of the Commons," *Science* 162 (1968).
9 Hansen, *Storms of My Grandchildren*, 189.
10 www.ipcc.ch/sr15/chapter/2-0/
11 Singer, *One World*, 48; Baer, "Equity, Greenhouse Gas Emissions, and Global Common Resources"; Jamieson, "Climate Change and Global Environmental Justice."
12 Jamieson, "Climate Change and Global Environmental Justice," 301; Singer, *One World*, 40.
13 Garnaut, *The Climate Change Review. Final Report*, 204.
14 Baer, "Equity, Greenhouse Gas Emissions, and Global Common Resources."
15 John Rawls, *A Theory of Justice* (Oxford: Oxford University Press, 1973), 60–80.
16 http://webarchive.nationalarchives.gov.uk/+/www.hm-treasury.gov.uk/stern_review_report.htm.

References

Baer, Paul. "Equity, Greenhouse Gas Emissions, and Global Common Resources." In *Climate Change Policy: A Survey*, edited by Stephen H. Schneider, Armin Rosencranz, and John O. Niles. Washington, DC: Island Press, 2002.
Baer, Paul, Tom Athanasiou, Sivan Kartha, and Eric Kemp-Benedict. "Greenhouse Development Rights: A Proposal for a Fair Climate Treaty." *Ethics Place and Environment* 12, no. 3 (2009): 267–81.

Gardiner, Stephen M. "Ethics and Global Climate Change." *Ethics: An International Journal of Social, Political, and Legal Philosophy* 114, no. 3 (2004): 555–600.

Gardiner, Stephen M. "The Global Warming Tragedy and the Dangerous Illusion of the Kyoto Protocol." *Ethics and International Affairs* 18, no. 1 (2004): 23–39.

Gardiner, Stephen M. "A Perfect Moral Storm: Climate Change, Intergenerational Ethics and the Problem of Moral Corruption." *Environmental Values* 15 (2006): 397–413.

Gardiner, Stephen M. "The Real Tragedy of the Commons." *Philosophy and Public Affairs* 30, no. 4 (2001): 387–416.

Garnaut, Ross. *The Climate Change Review: Final Report.* Cambridge: Cambridge University Press, 2008.

Hansen, James. *Storms of My Grandchildren.* New York: Bloomsbury, 2009.

Hardin, Garret. "The Tragedy of the Commons." *Science* 162 (1968).

Helm, Carsten. "Distributive Justice in International Environmental Policy: Axiomatic Foundation and Exemplary Formulation." *Environmental Values* 10, no. 1 (2001): 5–18.

Jamieson, Dale. "Climate Change and Global Environmental Justice." In *Changing the Atmosphere: Expert Knowledge and Environmental Governance*, edited by Clark Miller and Paul Edwards, 287–307. Cambridge, MA: MIT, 2001.

Leopold, Aldo. *A Sand Country Almanac.* Oxford: Oxford University Press, 1966.

Rawls, John. *A Theory of Justice.* Oxford: Oxford University Press, 1973.

Singer, Peter. *One World.* Melbourne: Text, 2002.

10 Being human

An ecocentric approach to climate ethics

Amanda M. Nichols

Introduction

"I am a human being." For those of us who are human, this statement has been fundamental in understanding our historical evolution, navigating our biocultural lifeworlds, and negotiating our place in the universe. But what does it actually mean to *be* "human"? And what does being human mean in a world replete with non-human organisms in an age of unprecedented climatic change?

Many ethical approaches dealing with the effects of anthropogenic climate change have been based on anthropocentric presuppositions influenced by or grounded in religion. These regressive religious and cultural narratives privilege human interest and minimize human obligations to our earthly co-inhabitants.

In increasing numbers, scholars in the humanities and social sciences are becoming concerned with climate change. Their approaches to dealing with this global problem tend to fall somewhere on the continuum between anthropocentric and ecocentric. Within the last decade, philosophical posthumanism[1] has been one such field that has taken a turn toward environmental concern.

Herein, I examine the ethical approach of ecocentrism alongside recent posthumanist scholarship in order to determine how, and if so to what extent, future dialogue between the two might stimulate mutually beneficial learning and contribute to the formulation of a practical non-anthropocentric climate ethic. Although the field of posthumanism was not founded with deep ecological or ecocentric values, recent approaches have sometimes moved dramatically in an ecocentric direction. Some scholars, including Donna Haraway and Edwardo Kohn, have made a profound move in that direction, advocating ecocentric environmental views that may help us deal with the complexity of living on a rapidly changing planet. Others, including Timothy Morton, remain confused, oscillating between repetition of a learned set of non-anthropocentric theories and a regression to anthropocentric ideologies issued as directives. Looking at the recent works of these three posthumanist scholars, I shall review the range of philosophical change taking place in posthumanism in light of global climate change.

Being human: what has it meant?

A worldview in which humans are granted superiority over all other beings proliferates. Human needs are judged as more important than those of non-humans, and human desires are often met at the expense of other species' wants, needs, and lives. Planetary needs are addressed by humans, from a human-centered perspective. The resulting speciesism, or human chauvinism, has already driven countless species to the brink of extinction—and beyond. This is the world in which we live today. Many scholars have attributed the mind-set of human superiority to the perpetuation of a particular anthropocentric worldview: Western Christianity. Perhaps best known among those was Lynn White Jr., who published his article "The Historical Roots of Our Ecologic Crisis" in 1967. Therein White argued that Western Christianity "is the most anthropocentric religion the world has seen" (White 1967, 1205) and is, at least in part, directly responsible for promoting the anthropocentric ideologies that have led to the destruction of our environment. Drawing on Biblical narratives to exemplify the anthropocentric and dualistic teachings he is critical of, White stated,

> God had created Adam and, as an afterthought, Eve to keep man from being lonely. Man named all the animals, thus establishing his dominance over them. God planned all of this explicitly for man's benefit and rule: no item in the physical creation had any purpose save to serve man's purposes. And, although man's body is made of clay, he is not simply part of nature: he is made in God's image.
>
> (White 1967, 1205)

White goes on to predict that these anthropocentric ideologies will exacerbate the "ecological crisis until we reject the Christian axiom that nature has no reason for existence save to serve man" (White 1967, 1207).

Bron Taylor has argued persuasively, however, that it was John Muir who developed the first explicit critique of religious ideas as promoting anthropocentric ideologies (B. Taylor 2016, 278). In his 1867 writing "Cedar Keys,"[2] Muir critiqued the Genesis story's claim that the world had been made for man, saying that it is pernicious and leads to an anthropocentric worldview that is detrimental to the environment (B. Taylor 2016, 278). Rachel Carson (1962) and Aldo Leopold (1949) also cited Christianity as being "complicit" in the environmental destruction that they evidenced in their writings. Moreover, both Clarence Glacken (1967) and Donald Worster (1977), in addition to White, argued that Western understandings of science and technology perpetuated problematic ideas about human relationships with the natural world and were largely influenced by the Christian worldview. More recently, in his 2015 Papal Encyclical, *Laudato Si: On Care for Our Common Home*, Pope Francis mirrored these sentiments, admitting that

> Christians have not always appropriated and developed the spiritual treasures bestowed by God upon the Church, where the life of the spirit is not

dissociated from the body or from nature or from worldly realities, but lived in and with them, in communion with all that surrounds us.

(Pope Francis 2015, 105)

In *Respect for Nature*, Paul Taylor expanded on the ideas of Muir, Leopold, and others like them and argued that since the European Middle Ages (roughly the fifth to 15th centuries C.E.), a worldview based on the Great Chain of Being proliferated in Western culture. The Great Chain of Being is a hierarchical ranking of the Earth's species, based on the anthropocentric Abrahamic religious stories described in the Old Testament of the Bible.[3] He wrote,

> The Great Chain of Being is the view that every existent thing has a certain place in an infinite hierarchy of entities extending from the most real and perfect to the least real and most imperfect. Beginning with God at the top, this hierarchal order continues down through various levels . . . to humans, followed by animals and plants . . . This is a metaphysical or ontological order as well as a valuational one. All things fall into a continuum of degrees of inherent worth that reflects the very structure of reality.
>
> (P. Taylor 1986, 139)

This hierarchical categorization has proliferated in Western culture and influenced ideas of what it means to be human in relation to the non-human world.

Humanism has also been problematic because it has linked rationality and the ability to reason to humanness. Taylor argued that this view stems from the ancient Greeks, including Pythagoras and Aristotle,[4] who deemed reasoning as superior to acting on the animalistic side of our nature by pursuing our passions and desires. Summarizing a Greek humanist attitude, Taylor stated,

> It is in living a fully rational life that we realize our truly human potentialities and thereby achieve our true good. Human happiness or well-being is the life of reason, and to guide our conduct by reason, both in the choice of means and of ends, is to exist on the highest level possible for a living thing.
>
> (P. Taylor 1986, 136)

The view that there is something that sets humans apart from other species was later adopted and expanded by René Descartes (1596–1650). Although Descartes believed that the mind (or soul) and body were distinct entities in humans,[5] he denied the existence of the mind in animals.[6] This contributed to a dualistic worldview, not only privileging the rational mind above the mechanical body, but also privileging humans over other species. Some feminist scholars[7] have argued that these dualistic understandings have also contributed to problematic dualistic understandings of gender, race, and sexuality. Many historians, ecofeminists, and others have critiqued this worldview, arguing that it marginalizes, or completely leaves out, certain groups that do not fit into binaries constructed in these ways (Merchant 1980). Ecofeminists, for example, have argued that categorical division is arbitrary because gender is performative and race and sexuality exist on a spectrum.[8]

To be human then has meant being superior—divinely, as we see represented in the Great Chain of Being—but also intellectually, as we see in Descartes, and for some, evolutionarily. These ideological views have pervaded Western philosophy, Western science, and Western society for generations. But as we enter an era of un-precedented anthropogenic climatic change, we must come to terms with our role in the destruction. We must also acknowledge and accept that humans are complicit in the death and destruction of countless other species. The perpetuation of anthropocentric worldviews continues to justify callous indifference to scores of living things. Ecocentrism, however, promises something different.

Ecocentrism explained

Ecocentrism is a non-anthropocentric ethic that extends ethical consideration to all components of the Earth's biotic and abiotic community. Although there is no universally agreed upon definition of ecocentrism, the following criteria are generally acknowledged as being necessary for an ecocentric worldview.[9] First, ecocentrists reject anthropocentrism and value non-human nature independently of any perceived or actual benefit or usefulness it might have to, and for, human beings. Ecocentrists adopt a biological sciences understanding of all species, including humans, as the products of long evolutionary processes that are still emerging, and they acknowledge that humans are relative newcomers to life on planet Earth. They are generally critical of many aspects of modern societies, including pro-growth economies (capitalist and socialist) that look at the world as a "resource" for humans, as well as consumerism and industrialism. Moreover, they acknowledge the need to lower the global human population for the preservation and sustainability of all life on Earth. Finally, ecocentrists are also focused on holistic entities (like ecosystems and biospheres) and argue that these should be granted ethical consideration and, in hard cases, even greater protective measures than individual parts or members (Ecocentric Alliance).

One of the leading 20th-century proponents of such a value orientation was Aldo Leopold. In his 1949 essay "The Land Ethic," Leopold wrote that "A thing is right when it tends to preserve the integrity, stability, and beauty of the biotic community. It is wrong when it tends otherwise" (Leopold 1949, 224–25). Leopold further developed this understanding when he stated

> the land ethic simply enlarges the boundaries of the community to include soils, waters, plants, and animals . . . A land ethic of course cannot prevent the alteration, management and use of these resources, but it does affirm their right to continued existence, and, at least in spots, their continued existence in a natural state.
>
> (Leopold 1949, 204)

Leopold also said that the land ethic changes the role of human beings, making us members, as opposed to conquerors of it (Leopold 1949, 204). Other scholars have suggested, though, that ecocentrism as an ethical system has evolved alongside humanity, existing in our consciousness and our interactions with the natural world. Washington et al. (2017) argued that "lore and law" from a variety of indigenous groups from around the world reflects a longstanding "ecocentric view of the world" (Washington et al. 2017, 35). It would be difficult, if not impossible, to conclusively argue that our ancestors lived by an ecocentric worldview. But there are ecocentric ethics apparent in some of our current ethical systems, as is illustrated in the following.

Deep ecology, an ethical system first advanced by Norwegian philosopher Arne Naess in 1973, provides another example of an ecocentric ethic. Naess's eight-point environmental ethic adheres to the ecocentric model presented earlier: Naess rejected anthropocentrism, called for the "flourishing of the living Earth" and demanded an ideological change and a "substantial decrease" in human population.[10] Although Naess has been critiqued as misanthropic,[11] many scholars and environmentalists have lauded his efforts. For example, the Ecocentric Alliance (EA), which is "a global advocacy network for ecocentrism and deep green ethics", has adapted Naess's deep ecological ethic as part of their own approach (Ecocentric Alliance). The EA was originally developed by David Orton in the early 1980s, who called it "left-biocentrism" (Ecocentric Alliance). Drawing on a biocentric ethic, Orton went further by adamantly rejecting capitalism and incorporating social justice. The resulting group promotes action and moral guidance to aid in "reversing the decline of non-human nature and on building economic systems and communities that are in harmony with the Ecosphere" (Ecocentric Alliance). Alongside the 11 principles from the Manifesto for Earth, Naess's eight-fold deep ecological ethic provides the framework for EA's agreement to preserve and protect the variety and species diversity of life on this planet.

The *Ecological Citizen*, an online, peer-reviewed journal that was founded in 2017 has also promoted the adoption of an ecocentric worldview among the scientific and environmental communities. Its mission states that

> Creating a harmonious, respectful and mutually flourishing relationship with the ecosphere is the basis of [an ecological] civilization. This involves preserving and restoring biological richness, ecological complexity and evolutionary potential—as well as the beauty, mystery and integrity of Earth. Nothing less will suffice. We are now looking global environmental collapse in the face, as our actions tear into the natural ecosystems that sustain all life, including our own, and inflict untold suffering on our fellow creatures. . . . Rather than dominating and parasitizing the biosphere, with non-human life harmed and ever-increasingly hemmed in by humans' industrial development, an ecological civilization would thrive within a preserved and restored expanse of unfragmented wild nature.
>
> (Ecological Citizen 2017)

Scholars from various fields have contributed to the journal, and it is apparent that a shift toward an ecocentric worldview is already well underway in parts of the academic community.

But where does posthumanism fall in this discussion? As a field that has often been considered on the forefront of the movement to think "beyond the human," crafting riveting tales of artificial intelligence and cyborgs, one might think that posthumanists would be leading the charge of thinking about the "beyond the human" environment. But that has not been the case. Posthumanists have, in fact and in general, been latecomers to the discussion of anthropogenic ecological destruction and climate change. Many who have joined the fervor have made huge strides across the spectrum from non-anthropocentrism, and some have turned directly to ecocentrism. With that in mind, we may question what affinities posthumanism has, or might have, with ecocentrism and whether posthumanism as a whole has shifted, or is in the process of shifting, its ethical stance.

Post-humanism: an ecocentric ethic?

Beginning in the late 20th century, philosophical post-humanism developed as one response to humanism, a philosophical perspective that values the human as an active agent and generally denies theistic worldviews in favor of scientific ones. Critical philosophical posthumanism differs markedly from other forms of post-humanism, including cultural posthumanism, transhumanism, and anti-humanism. In general, the philosophical form of post-humanism views the human as a biocultural creature and rejects established cultural legitimations of anthropocentric dominance. Moreover, it transcends the models of dualistic thinking proposed by Descartes and places human intelligence and rights on a spectrum with other-than-humans (including animal, plant, and cyborg). It expands common understandings of moral inclusion and recognizes the rights of all other species. It looks beyond the human to reconceive notions of intelligence, consciousness, cooperation, sentience, and emotion.

The growing recognition that we are living on a rapidly changing planet and already experiencing dramatic climatic shifts has encouraged some posthumanists to develop new theories about how humans view the world and their place in it, and some of these have advanced philosophical theories which promote ecocentric attitudes. Herein, I examine three scholars, whose posthumanist work has taken the ecological turn. Examining their writings alongside an ecocentric model, I show that posthumanism as a field is changing in light of ecological concern. Some scholars, like Donna Haraway and Edwardo Kohn, have moved dramatically toward extending ethical inclusion and promoting ecocentric worldviews. Yet others, such as Timothy Morton, cling to certain theoretical dogmas (some of their own creation) which hinder them from validating or adopting an ecocentric worldview.

Donna Haraway has come a long way from cyborgs and simians (Haraway 1991). Her theoretical views have, in fact, shifted so far from her early posthuman writings that she now tends to distance herself from that very field, finding it too anthropocentric and unwilling to take the value of non-human nature seriously. However, her work is generally still classified as philosophical posthumanism. In her two most recent works, *When Species Meet* (2008) and *Staying with the Trouble* (2016), Haraway has sought to broaden our knowledge of interspecies relationships and ecological responsibility. Upon examination of her work, it is apparent that Haraway's own ethical worldview has changed over time. Moreover, like Leopold, Haraway has begun to see the necessity of widening the circle of ethical inclusion to all members of the biotic and abiotic communities.

As early as 2008, Haraway had begun to shift toward an ecocentric worldview. In *When Species Meet*, Haraway discussed inter-species relationships looking specifically at animal ethics and the knowledge that can come from human and animal interaction. Therein, she argued that the use of other-than-human animals for human purposes deprives animals of any rights of their own. Bodies of other-than-humans are acted upon, she argued, in a manner that actively ignores social ques that transcend the species barrier (Haraway 2008, 24). Here, we can see that Haraway values non-human nature independently of the value it has to humans. Haraway also adamantly critiqued the capitalistic system of modern culture for its tendency to erase animal bodies as contributors to their biological communities. Capitalism, she argued, only represents human labor, leaving out the contribution of other-than-humans to that system. She continued, saying that Marx was "unable to escape from the humanist teleology of that labor—the making of man himself" and that his theory is problematic because "no companion species, reciprocal inductions, or multispecies epigenetics are in his story" (Haraway 2008, 46). The erasure of other-than-human bodies in the capitalistic system stems from a complex form of alienation—the incapability of consumers in the capitalist system to understand commodities as the products of natural and social relationships. Here, it becomes apparent that Haraway has not only rejected anthropocentric worldviews but has also rejected modern capitalism and consumerism.

Haraway's more recent book, *Staying with the Trouble* expanded on these ideas and showed that she has now moved even further along the spectrum toward an ecocentric worldview. Haraway argued that the ascribed ages that we, humans, are said to be embedded within (the Anthropocene and the Capitalocene) have created a problematic ideological understanding about our interconnections with other-than-human beings. Instead, while she claims that we have left the Anthropocene behind, she believes that pushing humanity forward into the age of the Capitalocene deprives us of a critical liminal moment in which we have the opportunity to gain an increasing awareness about our interactions with, and impact upon, the other-than-human species that share our world. This liminal age is the Chthulucene, which she described as "a kind of timeplace for learning to stay with the trouble of living and dying in response-ability on a damaged earth" (Haraway 2016, 2). Within this timeplace, Haraway argued, we must take on the task of "making kin" by learning to "live and die well with each other in a thick present" and determining "to whom one is actually responsible" (Haraway 2016, 1–2). She further stated,

> the Chthulucene is made up of ongoing multispecies stories and practices of becoming-with in times that remain at stake, in precarious times, in which the world is not finished and the sky has not fallen – yet. We are at stake to each other. Unlike the dominant dramas of the Anthropocene and Capitalocene discourse, human beings are not the only important actors in the Chthulucene, with all other beings able to simply react.
>
> (Haraway 2016, 55)

Here, we see similar ideas as those Haraway advanced in 2008—namely, the valuing of non-human nature independently of humans and rejection of anthropocentric worldviews and the capitalist system.

But Haraway goes further in this book. Adopting understandings from the biological sciences, she draws on some of the latest eco-evolutionary developmental biology[12] to advance a holistic evolutionary worldview. These sciences show that not only do all species co-exist interdependently but that evolutionarily even prokaryotic organisms are engaged in symbiotic relationships, which allowed for the formation of the first eukaryotic cells. She further stated that "critters – human and not – become-with each other, compose and decompose each other, in every scale and register of time and stuff in sympoietic tangling, in ecological evolutionary developmental earthly worlding and unworlding" (Haraway 2016, 97). This theory, known as symbiogenesis was substantiated by Lynn Margulis in 1967. Significantly, this theory ties together evolutionary biology with holism, by showing that all of the Earth's biological systems are interrelated. Nancy Moran elaborated, saying that this gives us an "enormous new ability to discover symbiont diversity, and more significantly, to reveal how microbal metabolic capabilities contribute to the function of hosts and biological communities" (Haraway 2016, 67). In short, Haraway is proposing a holistic worldview that acknowledges that biological systems are made up of interdependent entities, each of which plays a crucial role in the function of the entire system.

Haraway also tackles the issue of global human population in her chapter on "Making Kin." Kin, she said, is a word which originally meant "logical relations" in British English, but the meaning was shifted to "family members" in the 17th century (Haraway 2016, 103). She defined making kin as "making persons, not necessarily as individuals or as humans" and declared her hope for decreased human populations in the future (Haraway 2016, 103). She stated,

> Over a couple hundred years from now, maybe the human people of this planet can again be numbered 2 or 3 billion or so, while all along the way being part of increasing well-being for diverse human beings and other critters as means and not just ends. So make kin, not babies! It matters how kin generate kin.
>
> (Haraway 2016, 103)

Haraway's views concerning anthropocentrism, non-human nature, capitalism, evolutionary biology, and human population engage with and reflect ecocentric attitudes toward the non-human world and promote an ecocentric ethic to help us think about non-humans in an age of overwhelming climatic change.

Haraway is not the only posthumanist to advance similar perspectives. Anthropologist Edwardo Kohn has also made the turn to ecocentrism in his book *How Forests Think: Toward an Anthropology beyond the Human* (2013). Kohn's ethnographic study looked at human nature interrelationships in the Runa village in Ávila, Ecuador. Posing the critical question of "what would it mean to say that forests think," Kohn challenged categorical assumptions of what it "means to be human" that are reinforced by framing and representation (Kohn 2013, 7). He acknowledged that non-anthropocentric and holistic mentalities have been deepseated in cultures and traditions far beyond ecocentrism and posthumanism. Kohn has also moved quite far on the continuum toward ecocentrism, rejecting human exceptionalism and adopting holism.

Kohn argued that human understandings of representation are conventionally limited to those which are symbolic and linguistic and that social theory problematically "conflates representation with language" (Kohn 2013, 8). This, he says, implies that animals, and all forms of non-human nature, are incapable of representing themselves and reflecting on their state of being in the world: a supposition he finds fundamentally incorrect. "Along with finitude," he said, "what we share with jaguars and other living selves, whether bacterial, floral, fungal, or animal, is the fact that how we represent the world around us is in some way or another constitutive of our being" (Kohn 2013, 10). Here, Kohn not only challenged the idea of human exceptionalism in the ability to self-represent and self-reflect, but he also questioned what it means to be alive and to think in a way that helps us reimagine our relationality with other beings. He challenged readers to decolonize their thinking in order that they may better see the "moral webs that humans spin" and the network of relationships between humans and animals (Kohn 2013, 135).

Moreover, Kohn understands evolutionary biology as a given, noting the adaptive differentiation of the "proto-crabs" legs, "which allowed the organism as a whole to better fit or represent its environment" (Kohn 2013, 98) and that "walking stick lineages that were least noticed survived" and "came to propagate effortlessly into the future" (Kohn 2013, 176). For Kohn, humans are just one of the many relational emergent forms[13] that make up the world. Furthermore, Kohn discussed the role of death in the continued evolution of life on Earth. He noted that death is a "quality of being in future, which captures the logic of life's continuity and how this continuity is made possible thanks to the room each of our individual deaths can make for the lives of others" (Kohn 2013, 222).

Holistic worldviews are also important for Kohn. He stated that "we live in sociocultural worlds – complex wholes – that, despite their holism, are also open to that which lies beyond them" (Kohn 2013, 223). Here he is talking about representation, ways of knowing, and more specifically, the way we humans currently understand the world and other beings in it. What he means here is that a holistic worldview allows us to understand the world as a complex and interconnected system. Moreover, it also helps us to be open-minded about things that we do not yet know about other species and about planetary life systems. Ecocentrism, for Kohn, has contributed to increased understanding of the other, that which "looks back" at us as we look at it (Kohn, 92).

Philosopher and posthumanist scholar Timothy Morton has also recently turned toward approaches that deal with ecological concern. Several of his more recent works discuss climate change, co-existence, and the impact and effects of being human on a rapidly changing planet. However, he more than any of the others, holds back from fully adopting an ecocentric perspective.

Morton does advance many of the same views as ecocentrists. For example, he extended ethical consideration to all species, biotic and abiotic, and said they are "discrete entities with a 'life' of their own no matter whether a (human) subject has opened the epistemological refrigerator door to check them" (Morton 2016, 14). Morton also drew on Darwinian science to talk about the long evolutionary processes behind current species. He said that "contemporary science allows us to think species not as absolutely non-existent, but as floating, spectral entities that are not directly, constantly present" (Morton 2016, 18). This science also, as Morton stated, allows us to understand the evolutionary interconnections between all species. Additionally, Morton rejects anthropocentrism and capitalism and argued that "agents can't be reduced to their merely human use or exchange value" (Morton 2016, 21). He further critiqued Marx for failing to think of species (beyond the human) as creating their own environments (Morton 2016, 34) and said that Marxism has "imposed a host of inhibiting blocks to thinking the non-human" (Morton 2016, 24). To this point, Morton's views line up pretty well with an ecocentric ethic. But what about when it comes to holism?

Thanks to the efforts of Stewart Brand, who helped to publicize the first picture of the whole Earth taken from the moon in 1966, humans have been able to visualize "Earth" since the 1970s. Timothy Morton would argue, however, that this has brought us no closer to understanding it. Earth can be considered what Morton has called a hyperobject: a thing "massively distributed in time and space relative to humans" that is "viscous," "nonlocal," and "interobjective" (Morton 2010, 1). But what does that actually mean? In short, as best I can tell given his regrettably obtuse prose, Morton appears to imply that Earth is a holistic system that humans fail to fully understand. Because the Earth is so big, we humans are only ever capable of interacting with certain local manifestations of it. Moreover, it exists on a transhistorical time scale that we have difficulty quantifying. I surmise that hyperobjectivity, for Morton, also means that we have trouble conceptualizing all of the interconnected parts of the Earth system, from the role each individual plays to how ecosystems interact with one another. From this, it seems that Morton would advocate ethical holism, in order that we might be able to better understand the Earth's hyperobjectivity.

Morton rejects the idea of holism outright, however, saying that

> because there is thus no top object that gives all objects value and meaning, and no bottom object to which they can be reduced . . . it means that we

have . . . more parts than there are wholes. . . [which] makes holism of any kind totally impossible.

(Morton 2013, 116)[14]

Morton clings to this idea in his more recent book *Dark Ecology*, in which he stated "holism in which the whole is greater than sum of its parts depends on some (false) concept of smooth, homogenous universality or space or infinity" (Morton 2016, 12). He goes on to assert that "there never was a constantly present, easy to identify whole, because there was never a general, homogenous space box" (Morton 2016, 12). Here, Morton is describing Graham Harman's metaphysical, Heideggerian theory of object-oriented ontology (OOO).[15] This theory conceives of a world of objects—objects which exist apart from their relationships with humans and with other objects (Harman 2002). In OOO, relationships between objects, including relationships between human objects and non-human objects, as well as relationships between two non-human objects, leads to reification. Morton said, "*reification* is precisely *the reduction of a real object to its sensual appearance-for another object*" (Morton 2013, 119). In short, Morton is arguing that relationships between objects create distortion of some essential "objectness." Or, more clearly, "humankind exists, and I am a member of humankind. But there is so much more to me than being a member of humankind" (Morton 2017, 122).

In this volume, Sam Mickey has detailed how Morton's conceptualization of OOO leads to his call for "ecology without nature"[16] (Morton 2013, 121) and a recognition of the differences between explosive and implosive holism (Mickey 2019, 7–8). Morton critiqued explosive holism for its implied value judgment that "the whole is especially different (better or worse) than the part" (Morton 2017, 121). He goes on to argue in favor of implosive holism, an idea that leads him to the conclusion that "nothing causes global warming" because it is "heaps" of things (individuals, actions, and so forth) combined together that manifest as global warming so there can be no identifiable "cause" per se (Morton 2017, 122–23).

Herein, I see two main issues with the line of reasoning that Morton has advanced across these four volumes. First, he has constructed an artificial dualism between humans and nature, which retains the anthropocentric understanding that humans are something set apart and distinct from the natural world. Humans are not apart from nature biologically, nor can we be so easily absolved of our shared responsibility with and to the rest of the natural world. Moreover, while he adamantly disavows any idealized view of nature, Morton simultaneously avers that there is a "proper relationship with the earth and its lifeforms, which would, of course, include ethics and science" (Morton 2007, 2). The logical incongruity and intellectual doublespeak is unsurprising.

Second, Morton's arguments are dependent (pun intended) on a causal ontology adapted from Buddhism, a fact that he never directly acknowledges but alludes to in all four of the works in question. Morton's understanding of the doctrine of co-dependent origination, or *paṭiccasamuppāda*,[17] however, is, at best, limited. First, Morton claims that in the causal chain of co-origination, "the whole is less than the sum of its parts" implying a value judgment. But according to the story of King Milinda and the chariot in the Sutta Pittaka,[18] there is not value placed on any part of the causal chain over any other part, nor any individual part of the culmination of those parts. Second, Morton's argument regarding OOO, that the relationships between objects creates reification and therefore detracts from some essential "objectness" is also misguided according to a Buddhist perspective. The very point of the doctrine of dependent origination is that nothing exists apart from the embedded relationships they share within causal chains.

This is where Morton's views are most clearly problematic. He fails to see that we, humans, are not only inextricably embedded within these relationships, a fact that Haraway and Kohn both recognize, but he also fails to see that it is these very relationships that will help us to relate and to reject dualistic ideologies of racism and speciesm, which is what he claims to want.[19] Morton's theory is also problematic because by claiming, for instance, "that nothing causes global warming," he absolves individuals and our species as a whole of responsibility, allowing us to throw our hands up in the air in frustration and look for results from elsewhere— which is exactly what Morton seems to have done. Indeed, there is an implicit dismissal of science in his failure to acknowledge that geologists are on the verge of declaring the Anthropocene as the most recent geological era, thus declaring that humans and the "dominant animal mentality" have decisively contributed to and precipitated the current changes to the climate.

Morton concluded *Hyperobjects* with Heidegger's line "only a god can save us now" but added "we just don't know what sort of god" (Morton 2010, 201). This is a remarkable abdication of responsibility.

As we have seen, the reliance on this mentality is part of what has gotten us here in the first place. Gods cannot save us now. And as for Morton, he fails to provide any type of practical application for dealing with the problem of anthropocentric climate change.[20] In his recent review of Morton's *Being Ecological* (2018), Adam Dickerson also expressed this sentiment stating,

> Most crucially, the book has nothing to say about what ethics and politics might follow from this nonanthropocentric awareness. Morton is systematically evasive when it comes to saying something about ethical principles or resolving ethical problems – let alone addressing questions about political strategy, concrete actions, policies, organizational structures and so forth.
>
> (Dickerson 2019)

Overall, Morton's ethical attitudes, which may have seemed as though they were moving toward ecocentrism, are still problematically ensconced in anthropocentrism.

Toward posthuman ecocentrism

Many scholars and scientists have made claims to the effect that the "natural state" envisioned by Leopold (204) has come to an end and that we are now living in a post-natural world.[21] Bill McKibben, for example, spoke of the "end of nature," which Holmes Rolston picked up on when asserting in this volume that "nature is over."[22] But nature is not some idealized state, Platonic form, or paradise that we have lost. What we call "nature" is the product of long evolutionary bio-cultural process. Indeed, modern humans have left their mark, manipulating our physical environments and altering the pace of climatic change. But nature has not and will not cease to exist. While humans may be able to alter the course of evolutionary processes to some degree, to stop them entirely—to stop nature from existing entirely—is, I surely hope, well beyond our capacity. And to think that nature will (or has already!) cease to exist because of humans is not only exceptionally naïve but also troublingly anthropocentric. It also fails, fundamentally, to recognize that we humans are ourselves a part of nature, and if it is dead, then so must we be.

There have been significant changes in the field of philosophical posthumanism within the last decade. Some posthumanists have effectively decentered the human in ethical discussions and often argue that human rights, feelings, and lives should not be privileged over other species'. Drawing on Darwin's theory of biological evolution, posthumanists are beginning to understand the human animal as only one among many evolutionarily adapted and adapting beings. The field has taken a self-reflexive turn: it acknowledges human knowledge as only one way of knowing and understands that as the way humans make sense of the specific biocultural lifeworlds we inhabit. Subjectivity and moral concern are extended beyond the human species. Moreover, some of these newer forms of posthumanism challenge regressive religious and cultural narratives that privilege human interest. In light of the worsening impacts of climate change, posthumanists are one group that has begun to contemplate the possibility of a post-anthropocentric climate ethics. Based on the brief exploration I have done here, it seems that future dialogue between ecocentrists and posthumanists could yield fruitful results along such lines and hopefully facilitate direct action in a time, as Haraway said, that "remain[s] at stake, in precarious times, in which the world is not finished and the sky has not fallen – yet" (Haraway 2016, 55).

Notes

1 Here, I am referring to the branch of posthumanism known as "critical posthumanism" or philosophical posthumanism as defined by Parmod Nayar in *Posthumanism* (2014). Nayar stated "critical posthumanism seeks a more inclusive definition of the human and life and, for its theoretical-philosophical methodology, draws upon all those discourses, representations, theories and critiques of traditional humanism. . . [and] redefine[s] the boundaries of the human, and call[s] into question the hierarchies of human/non-human, human/machine and human/inhuman" (Nayar 2014, 6). Critical posthumanism rejects the anthropocentric assumptions of transhumanism, namely, the embrace of technological advancements meant to improve human intellectual and physiological capacities. It also rejects human exceptionalism and human instrumentalism or the idea that humans have the right to control the natural world.

2 Which was later published as the sixth chapter of Muir, John. *A Thousand Mile Walk to the Gulf.* (Boston and New York: Houghton Mifflin Company, 1916).

3 There are also some apparent links between how species are ranked hierarchically in the Great Chain of Being with Aristotle's biological ordering of species in his *History of Animals* (written in the 4th century BCE).

4 In *Metaphysics: Book 1, Part 5*, Aristotle mentions the Pythagorean table of opposites. This table, still largely in use during Aristotle's time by the Pythagoreans, listed ten (Pythagoras's perfect number) sets of opposite pairs. Aristotle stated, "Other members of this same school say there are ten principles, which they arrange in two columns of cognates – limit and unlimited, odd and even, one and plurality, right and left, male and female, resting and moving, straight and curved, light and darkness, good and bad, square and oblong. In this way Alcmaeon of Croton seems also to have conceived the matter, and either he got this view from them or they got it from him; for he expressed himself similarly to them. For he says most human affairs go in pairs, meaning not definite contrarieties such as the Pythagoreans speak of, but any chance contrarieties, e.g. white and black, sweet and bitter, good and bad, great and small. He threw out indefinite suggestions about the other contrarieties, but the Pythagoreans declared both how many and which their contrarieties are."

5 See Descartes, Rene. *Meditations VI*: "And although possibly (or rather certainly, as I shall say in a moment) I possess a body with which I am very intimately conjoined, yet because, on the one side, I have a clear and distinct idea of myself inasmuch as I am only a thinking and unextended thing, and as, on the other, I possess a distinct idea of body, inasmuch as it is only an extended and unthinking thing, it is certain that this I (that is to say, my soul by which I am what I am), is entirely and absolutely distinct from my body, and can exist without it."

6 See Rene Descartes. "Letter to More," *Replies to Objections IV*. February 5, 1649.

"Second, since art copies nature, and people can make various automata that move without thought, it seems reasonable that nature should produce its own automata, much more splendid than artificial ones – namely the animals. This is especially likely since we know no reason why thought should always accompany the sort of bodily structure that we find in animals. That no animal contains a mind isn't as astonishing as the fact that every human body contains one."

7 See, for instance, Stanton and Spalding (1885), Johnson (1992), Anderson and Clack (2004), Bynum, Harrell, and Richman (1986), Merchant (1980, 1996, 2007), Plumwood (1993), Ortner (1972).

8 See, for instance, Adams and Gruen (2014), Bauman (2018), Butler (1988), and Plumwood (1993).

9 See Curry (2011), *Ecocentric Alliance, and* Woodhouse (2018, *1–9, 96–106). For a more complete overview, see* Mosquin and Rowe (2004*).

10 Naess first presented these ideas in 1972, at his presentation at the Third World Future Research Conference in Bucharest.

11 See Bookchin (1987).

12 A subfield of biology that draws on Darwinian evolution to determine ancestral relationships and the co-evolution of species. A recent example is The Great Tree of Life, a genetic mapping that shows that "all species are related via common descent and part of a great tree of all life [which] dates to the time of Darwin." For more information see Soltis and Soltis (2019). See also Haraway (2016, 5, 7, 97, 98, 122, 150).

13 A relational emergent form is "A phenomenon that cannot be reduced to the contingent histories that give [it] specific characteristics. . . . [It] is never fully separable from [that] from which it emerges. . . . And yet. . . [it] is something other than [that] which it requires" (Kohn 2013, 166). "Emergent phenomena, then, are nested. They enjoy a level of detachment from the lower order processes out of which they arise. And yet their existence is dependent on lower-order conditions. This goes in one direction . . ." (Kohn 2013, 167).

14 See also Morton (2017, 122).

15 See Morton (2013, 13–14, 18), Harman (2002, 16, 2011).
16 See also Morton (2007).
17 Sanskrit: परतीत्यसमुत्पाद pratītyasamutpāda; Pali: पटिच्चसमुप्पाद *paṭiccasamuppāda*.
18 Book Two Lakkhana Pañha, Milindapanha, Sutta Pittaka.
19 See Morton (2017, 122).
20 See also Dickerson (2019).
21 See Rolston's article in this volume where he sites McKibben, Soule, Wapner, and Meyer. See McKibben (1989), Soulé (1989), (Meyer 2006), (Wapner 2010). See also Earle (1995, 142) and Walsh (2012).
22 Rolston – See Rolston 's article in this volume (2020, 203).

References

Adams, Carol and Lori Gruen. *Ecofeminism: Feminist Interactions with Other Animals and the Earth*. New York and London: Bloomsbury, 2014.

Anderson, Pamela S. and Beverley Clack. *Feminist Philosophy of Religion: Critical Readings*. New York: Routledge, 2004.

Bauman, Whitney A. *Meaningful Flesh: Reflections on Religion and Nature for a Queer Planet*. Earth, Milky Way: Punctum Books, 2018.

Bookchin, Murray. "Social Ecology Versus Deep Ecology: A Challenge for the Ecology Movement". Newsletter of the *Green Program Project*. No. 4 and 5. 1987.

Butler, Judith. "Performative Acts and Gender Constitution: An Essay in Phenomenology and Gender Theory." *Theater Journal* 40, no. 4 (1988): 519–31.

Bynum, Caroline, Stevan Harrell, and Paula Richman. *Gender and Religion: On the Complexity of Symbols*. Boston, MA: Beacon Press, 1986.

Carson, Rachel. *Silent Spring*. Boston: Houghton Mifflin, 1962.

Curry, Patrick. *Ecological Ethics: An Introduction*, 68. Malden: Polity Press, 2011.

Descartes, Rene. "Letter to More." *Replies to Objections IV*. February 5, 1649.

Descartes, Rene. "Meditations VI." In *Great Books of the Western World: 31. Descartes— Spinoza*, edited by Mortimer J. Adler, 98. Chicago: University of Chicago Press, 1952.

Dickerson, Adam. "Damaging Thinking: A Review of Timothy Morton's *Being Ecological*". *The Ecological Citizen* 2, no. 2 (2019): 199.

Earle, Sylvia A. *Sea Change: A Message of the Oceans*, 142. New York: Random House Publishing, 1995.

Ecocentric Alliance. https://ecocentricalliance.org/.

Francis, Pope, I. *Laudato Si': On Care for Our Common Home*, 105. United States Conference of Catholic Bishops, 2015.

Glacken, Clarence J. *Traces on the Rhodian Shore: Nature and Culture in Western Thought from Ancient Times to the End of the Eighteenth Century*. Berkeley: University of California Press, 1967.

Haraway, Donna. *Simians, Cyborgs, and Women: The Reinvention of Nature*. New York and Abingdon: Routledge, 1991.

Haraway, Donna. *Staying with the Trouble: Making Kin in the Chthulucene*. Durham: Duke University Press, 2016.

Haraway, Donna. *When Species Meet*. Minneapolis: University of Minnesota Press, 2008.

Harman, Graham. "Critical Animal with a Fun Little Post." *Object-Oriented Philosophy* (blog). October 27, 2011. https://doctorzamalek2.wordpress.com/2011/10/17/critical-animal-with-a-fun-little-post/.

Harman, Graham. *Tool-Being: Heidegger and the Metaphysics of Objects*, 16. Peru, IL: Open Court, 2002.

Johnson, Elizabeth. *She Who Is: The Mystery of God in Feminist Theological Discourse.* New York: The Crossroad Publishing Company, 1992.

Kohn, Eduardo. *How Forests Think: Toward an Anthropology Beyond the Human.* Berkeley: University of California Press, 2013.

Leopold, Aldo. "The Land Ethic." In *A Sand County Almanac*. Oxford: Oxford University Press, 1949.

McKibben, Bill. *The End of Nature*. New York: Random House, 1989.

Merchant, Carolyn. *American Environmental History: An Introduction*. New York: Columbia University Press, 2007.

Merchant, Carolyn. *The Death of Nature*: *Women, Ecology, and the Scientific Revolution.* New York: Harper San Francisco, 1980.

Merchant, Carolyn. *Earthcare: Women and the Environmental Movement*. New York and Abingdon: Routledge, 1996.

Meyer, Stephen M. *The End of the Wild*. Cambridge: The MIT Press, 2006.

Mickey, Sam. "Atmospheres of object-oriented ontology". *In Climate Change Ethics and the Non-Human World. Edited by Brian G. Henning and Zach Walsh* (Routledge 2020) pp. XXXX.

"Mission Statement." *The Ecological Citizen* (2019): 7–8. www.ecologicalcitizen.net/.

Morton, Timothy. *Being Ecological*. Cambridge: The MIT Press, 2018.

Morton, Timothy. *Dark Ecology: For a Logic of Future Coexistence*. New York: Columbia University Press, 2016.

Morton, Timothy. *The Ecological Thought*. Cambridge: Harvard University Press, 2010.

Morton, Timothy. *Ecology Without Nature: Rethinking Environmental Aesthetics.* Cambridge: Harvard University Press, 2007.

Morton, Timothy. *Humankind: Solidarity with Nonhuman People*. London: Verso, 2017.

Morton, Timothy. *Hyperobjects: Philosophy and Ecology After the End of the World*. Minneapolis: University of Minnesota Press, 2013.

Mosquin, Ted and Stan Rowe. "A Manifest for Earth." *Biodiversity* 5, no. 1 (2004): 3–9.

Muir, John. *A Thousand Mile Walk to the Gulf*. Boston and New York: Houghton Mifflin Company, 1916.

Nayar, Parmod K. *Posthumanism: Themes in 20th and 21st Century Literature and Culture*. Malden: Polity Press, 2014.

Ortner, Sherry B. "Is Female to Male as Nature Is to Culture?" *Feminist Studies* 1, no. 2 (1972): 5–31.

Plumwood, Val. *Feminism and the Mastery of Nature*. New York and Abingdon: Routledge, 1993.

Rolston, Holmes III. 'Wonderland Earth in the Anthropocene epoch'. *In Climate Change Ethics and the Non-Human World*. Edited by Brian G. Henning and Zach Walsh (Routledge 2020) pp. XXXX.

Soltis, Douglas E. and Pamela S. Soltis. *The Great Tree of Life*. Amsterdam: Elsevier Inc., 2019.

Soulé, Michael E. "Conservation Biology in the Twenty-First Century: Summary and Outlook." In *Conservation for the Twenty-First Century*, edited by David Western and Mary Pearl. Oxford: Oxford University Press, 1989.

Stanton and Spalding. "Has Christianity Benefitted Women?" *The North American Review* 140, no. 342 (May 1885): 389–410.

Taylor, Bron. "The Greening of Religion Hypothesis (Part One): From Lynn White, Jr and Claims That Religions Can Promote Environmentally Destructive Attitudes and Behaviors to Assertions They Are Becoming Environmentally Friendly." *Journal for the Study of Religion, Nature and Culture* 10, no. 3 (2016).

Taylor, Paul. *Respect for Nature: A Theory of Environmental Ethics*. Princeton: Princeton University Press, 1986.

Walsh, Bryan. "Nature Is Over: Little Is Left Untouched by Humans—And That Demands a Rethink of Environmentalism." *Time Magazine*. 2012. http://content.time.com/time/subscriber/article/0,33009,2108014-3,00.html.

Wapner, Paul. *Living Through the End of Nature: The Future of American Environmentalism*. Cambridge: MIT Press, 2010.

Washington, Hayden, Bron Taylor, Helen Kopnina, Paul Cryer, and John Piccolo. "Why Ecocentrism Is the Key Pathway to Sustainability." *Ecological Citizen* 1, no. 1 (2017): 35–41.

White, Lynn. "The Historical Roots of Our Ecologic Crisis." *Science* 155, no. 3767 (1967): 1203–7.

Woodhouse, Keith Makoto. *The Ecocentrists: A History of Radical Environmentalism*, 1–9, 96–100. New York: Columbia University Press, 2018.

Worster, Donald. *Nature's Economy: A History of Ecological Ideas*. Cambridge: Cambridge University Press, 1977.

11 Atmospheres of object-oriented ontology

Sam Mickey

Around a century ago, a debate was happening within the astronomical community as to whether there is more than one galaxy in the universe. In 1924, the answer was discovered when Edwin Hubble, using a high-powered telescope, found evidence that there are in fact galaxies beyond our own Milky Way, and by the end of the 20th century, as Brian Thomas Swimme and Mary Evelyn Tucker observe, "we discovered nearly a hundred billion galaxies. Each of these contains several billion stars. What does this mean for understanding our place amidst such vastness?"[1] The task of understanding the place of humankind has been further complicated by climate change and other environmental crises. Just as scientists are learning about the vast and complex universe of which we are a part, we are learning that human impacts on the Earth system are degrading and destroying the conditions of life on Earth, which includes the conditions of human life as well. What, then, is to be done? Centering our ethical concern primarily or exclusively on humans seems woefully inadequate for responding to the abundance and diversity of non-humans with which we coexist. Should humans direct ethical concern to all animals, all living organisms, whole ecosystems, the whole Earth, the whole universe? What about artificial objects, like human art, architecture, and technology? If ethics is not centered on the human (*anthropos*), where should it be centered? What if everything matters?

These kinds of questions come up in the documentary film, *Living in the Future's Past*, which features scientists, scholars, military leaders, politicians, and others who are committed to understanding and responding to the environmental challenges facing humankind. The narrator, Jeff Bridges, considers the following thought experiment: What if everything mattered equally? What if every single thing existed in the same way?

> Let's take a moment to image entering a theoretical world, where every *thing* in existence, at all scales, has equal value, from skyscrapers and trees, to humans, cola, cups, and orangutans, shoes and suitcases, vessels, water, diamonds, and dragonflies, even time itself. If we can imagine every *thing* as intrinsically equal, then we'd know that what we perceive as reality comes from the value judgments that exist in our minds.[2]

The thought experiment continues as the narrator's voice gives way to an ecological philosopher featured in the film, Timothy Morton, who puts the issue simply: "If everything exists in the same way, then we have an interesting problem." The problem is that, you still have to act and make choices, and some beings inevitably get ignored or harmed. If you love your cat, you are probably going to give less ethical priority to fleas or some such parasite feeding on your cat and even less priority to the potentially harmful bacteria the fleas carry. If you want to care for the whole system instead of focusing on individual organisms, you have to decide the scale of the system (e.g., ecosystem, bioregion, continent, Earth, or cosmos), and you still have to decide which parts of the system are integral to its functioning and which are harmful, so the problem of choosing between individuals remains. If everything has equal value, ethics is opened up far beyond the human, with no simple answer regarding what matters most.

This theoretical perspective, in which every being exists in the same way, with equal value, is what Morton calls object-oriented ontology, abbreviated as OOO (pronounced "triple-O"). In the mesh of entities described in Morton's version of OOO, every being (i.e., "object") has equal value, from molecules and fleas, to icebergs and polar bears, humans and cats, forests and cities, even a particularly massive object ("hyperobject") like global warming.[3] Anthropocentrism is too narrow to address multiple scales of beings that intersect in global warming. The predominant forms of non-anthropocentrism (i.e., biocentrism and ecocentrism) have not been very successful in handling global warming either, although they have contributed much toward thinking beyond the human scale.

Object-oriented ontology supports an alternative approach to climate ethics, a non-anthropocentrism of things themselves. Ontology—the branch of philosophy that studies "being" (*ontos*)—might sound like an intellectual distraction from urgent problems. Perhaps now is not a time for theory but for action. For Morton, thinking about the way things are really matters. Different ways of thinking involve different affective ways of being, different moods and styles, which are more or less conducive to sustainable coexistence with non-humans.

Object-oriented climate ethics extends the notion of "mood" or "attunement" (*Stimmung*) articulated by the German philosophers Immanuel Kant and Martin Heidegger. Attunements are affective modes of comportment. In other words, attunements compose the attitudes through which humans engage reality. Simply put, they are ways of tuning into the world. "If we want a good reality—say, for instance, non-violent coexistence between all beings—we might need to figure out what kinds of attitude are conducive to such a reality."[4] What moods facilitate sustainable or nonviolent coexistence with the myriad entities imbricated across multiple scales in climate change? Before considering the question of attunement, it is important to elaborate on Morton's ecological rendering of object-oriented ontology, which can be described as a kind of multicentrism, and to indicate the place of hyperobjects therein.

Multicentrism

Ecology is about interconnectedness, what Morton calls *the mesh*, but it is also about the entities that are interconnected, Morton's *strange strangers*.[5] While interconnectedness is fundamental to ecological thought, the entities that are interconnected are not thereby reducible to the mesh. There are beings (things, objects, entities) and relations (interactions, interconnections, symbioses). Strange strangers are not reducible to their relations or to their constituent parts and processes. Describing beings as strange strangers connotes the otherness that compels ethics beyond egoism into altruism. Such otherness or "alterity" is something emphasized by the French philosopher Emmanuel Levinas, whose work can be described as "ethical metaphysics"—a theory for which the study of reality (metaphysics) is a study of the ethical value in the otherness of the real, specifically the otherness of human beings.[6] Morton's strange strangers include all beings, not just humans. Furthermore, strangeness is not simply other. It can be found even in beings with which one is most familiar and most intimate, including companions, neighbors, and oneself.

Morton inherits his view from a French philosopher deeply influenced by Levinas, Jacques Derrida, who integrates the ethics of otherness with Heidegger's philosophy, "called *Destruktion* ('de-structuring'), and which Derrida calls deconstruction"—one of Morton's "favorite philosophical regions."[7] Thinking about deconstruction *and* ecology means thinking of deconstruction *as* ecology.[8] Deconstruction and ecology both track the ways in which things are not fully present substances, simply here and now, but are always entangled in contexts (deconstruction) or complex systems (ecology) that extend far beyond themselves. The ecological thought follows the deconstruction of the "metaphysics of presence," which, simply stated, is the metaphysical idea that something must be present for it to exist, like an independent and unchanging substance.[9] The term "strange stranger" is Morton's way of translating Derrida's *arrivant* ("newcomer," "guest"), which is someone whose arrival always remains "to come" (*à venir*), in the future (*l'avenir*).[10] When the other arrives, the other has baggage. Every guest carries unknown pasts, and thus the arrival of a guest is also the return of the past, such that the *arrivant* is a *revenant*. The guest is a ghost, a specter, one who returns from a long absence, and as a verb (the present participle of *revenir*, "to return"), the *revenant* is a movement of returning. Ontology is "hauntology" for Derrida, such that to exist is to be a haunting guest, a flickering of past and future, like a ghost seeking justice, a stranger seeking refuge, a newcomer seeking hospitality.[11]

The deconstruction of presence shows up in Heidegger's tool analysis in *Being and Time*, where he demonstrates the way in which being present-at-hand (*vorhanden*) depends on being ready-to-hand (*zuhanden*). A tool can become present, for instance, when it breaks and has to be fixed, but when everything is working, it must withdraw (*zurückzuziehen*) from any direct theoretical or practical concern to stay ready.[12] Morton puts it this way:

> Things are present to us when they stick out, when they are malfunctioning. You're running through the supermarket hell bent on finishing your shopping trip, when you slip on a slick part of the floor (someone used too much polish). As you slip embarrassingly toward the ground, you notice the floor for the first time, the color, the patterns, the material composition—even though

it was supporting you the whole time you were on your grocery mission. Being present is secondary to just sort of happening, which means, argues Heidegger, that *being isn't present*, which is why he calls his philosophy deconstruction or destructuring. What he is destructuring is the metaphysics of presence.[13]

The distinction between withdrawal and presence in the tool analysis is a crucial component of OOO, which Morton adapts from Graham Harman.[14]

For Harman, every object is a fourfold that is split along one axis into the withdrawal of the real and the relational interactions of the sensuous and along another axis between objects and their qualities, thus rendering a quadruple object (real object, real qualities, sensuous object, sensuous qualities).[15] Objects are enmeshed with other things, but in themselves, objects withdraw from presence, remaining strange, other, ungraspable—irreducible. Preserving the irreducibility of objects, OOO avoids two kinds of reductionism, "undermining" and "overmining," which can combine into "duomining."[16] Undermining reduces things to their constituent parts or to an underlying field, like claiming that a frog is really just atoms or energy, and the appearance of that energy as a frog to a human observer is due to the adaptive traits inherited in the human nervous system for responding to stimuli. Overmining reduces things to their effects or their relations with overarching systems, for instance, when a frog is viewed as a linguistic signifier ("frog") that reflects a socially constructed reality, and even atoms and energy are viewed as social constructs. For overmining and undermining, the frog is not real. It is either reduced downward to underlying material conditions or upward to human interpretation and social construction. In the case of duomining, it is reduced to some combination of the two (e.g., material processes that are socially constructed as a "frog"—undifferentiated energy that attains frog form through human language). OOO agrees that a frog has underlying material conditions, and that the meaning of a frog for humans depends partly on interpretation and construction, but OOO also argues that there is a real frog, a frog that is irreducibly itself and not merely an amalgam of underlying conditions and/or overlying interpretations.

Another reductionism rejected by OOO is anthropocentrism, specifically the idea that subjective capacities, including their normative implications, are only attributable to human beings. Theories of animal rights and biocentrism claim that forms of selfhood (subjectivity and self-organization) can be found in animals and even in all organisms. OOO makes a speculative claim that every object has some kind of subjectivity, some way of touching other entities and not merely being touched by them. In what Harman calls "vicarious causation," objects have mediated access to one another through their qualities, such that objects touch each other indirectly. They *"touch without touching."*[17] A human interprets a fly without ever fully exhausting its interpretive possibilities, and likewise, a frog interprets a fly without ever exhausting its interpretive possibilities. Analogously, a fire burns paper without having total access to the paper. Fire translates the paper's flammable qualities into the fire's terms, just like a frog translates a fly into the frog's world, and a human translates a frog into a human world. This means that there is some kind of agency or subjectivity in all things. There is some kind of translation, interpretation, or evaluation in all interactions; in other words, there is some kind of artistic or aesthetic dimension in all causality. "*The aesthetic dimension is the causal dimension.*"[18]

Undoing the privileged ontological status of humans, OOO is a "flat ontology."[19] It does not flatten out the differences between beings. A hawk is different from a human, but hawks, humans, and all things exist equally, each in the same way, as irreducibly unique beings (or whatever word we want to use for "beings": agents, artists, guests, ghosts, specters, strange strangers). OOO affirms the unique differences of all beings, and it flattens the walls that are used to categorically separate human beings from all other beings. Anthropocentrism erects walls between humans and non-humans, granting humans privileged access to properties like mind, soul, language, or rationality, and using those properties as grounds for excluding non-humans from ethical concern. As Amanda Nichols observes in the present volume, those anthropocentric walls are also how some humans attempt to gain power over other humans, as with sexist and racist systems for which women and people of color are not considered to be fully human but are seen as irrational, emotional, animalistic, or closer to nature (pp. 133–49). OOO joins historians, ecofeminists, and environmental justice advocates who have critically interrogated the intersections of racism, sexism, classism, and speciesism.

The flat field of OOO is a field in which all things, at all scales, exist equally. Everything is an agent, but nothing is an independent substance. Every agent comes with a compelling story, like a guest with a strange past and an unknown future, like a ghost demanding justice. Things are less like substances and more like specters. The flat field of OOO is "the *Spectral Plain*."[20] "There are some basic rules of politeness on the Spectral Plain, and these have to do with the idea of *hospitality to strangers*."[21] This ethics of hospitality is a radical altruism. Hospitality involves welcoming the stranger—the guest, the ghost—without appropriating or assimilating the stranger's radical otherness. It is not about tolerating others but genuinely affirming and welcoming others on their own terms. A phrase like "you are tolerated" sounds relatively hostile compared with the phrase "you are welcome."

For Derrida, "deconstruction is hospitality to the other."[22] Welcoming the other is an ongoing and ultimately impossible task, for the simple reason that welcoming one stranger often means not welcoming others, like welcoming my cat requires me to not welcome fleas. Furthermore, the very act of welcoming a stranger makes the stranger less strange. This is captured in the proverb, familiarity breeds contempt. Welcoming the other can harm the other, erasing or antagonizing the unique conditions of the other's existence. Hospitality requires that one pay attention to the complicity between being a host and being hostile, a complicity expressed in Derrida's portmanteau, "hostipitality" (*hostipitalité*).[23] Welcoming the other requires openness, vulnerability, and susceptibility; otherwise, the hostility entangled with hospitality damages the other, like confining a large wild animal to a small cage.

The rules of politeness on the Spectral Plain not only exceed the boundaries of anthropocentrism. They also exceed the boundaries of the predominant forms of non-anthropocentrism in environmental ethics, namely biocentrism and ecocentrism. Biocentrism and ecocentrism respectively take individual organisms and whole ecosystems as the primary locus of value. Biocentrism and ecocentrism are accurately described by Anthony Weston as "mega-centrisms."[24] They sort the profoundly diverse plurality of objects into categories of Life (biocentrism) and Land (ecocentrism). Biocentrism and ecocentrism are mutually exclusive. Biocentrism attributes value to organisms but not directly to nonliving entities or to whole ecosystems (Attfield). Ecocentrism attributes value to wholes but not directly to individual organisms (Nichols; Rolston). They each exclude some kinds of entities from ethical concern. A river does not have intrinsic value for biocentrism, although it matters indirectly because of all the living organisms in and around it. An individual salmon does not have intrinsic value for ecocentrism, although it has systemic value for the whole of which it is a part.

Ethics on the Spectral Plain is a radical pluralism that welcomes all strangers, at all scales, with attention to the unique conditions and demands of each, whether living, nonliving, or anything in between. This is not unlike Weston's multicentrism, for which "more-than-human others enter the moral realm on their own terms, rather than by expansion from a single center—a vision according to which there are *diverse* centers, shifting and overlapping but still each with its own irreducible and distinctive starting-point."[25] For Weston, a "multicentered ethic" means that "the growth of moral sensitivity and consideration does not proceed through an expanding series of concentric realms" but through experiencing "a world of separate though mutually implicated centers," an "*ex*-centric" world of "radical and polymorphous pluralism," a pluralism of "things themselves."[26] Applying this ex-centric ethics to anthropogenic climate change requires clarification of what climate is from an object-oriented perspective: a hyperobject.

Hyperobjects

In the same way that an atmosphere can refer to a mood (e.g., a party atmosphere) or to a system of gases surrounding a planet, climate has sociocultural and biophysical dimensions. Climate change involves a social system that, for the sake of progress and wealth accumulation, risks unprecedented scales of destructive change, shutting down the program of nature running in the background of culture, and adding so much carbon, plutonium, plastic, Styrofoam, and profuse artificial chemicals to Earth's crust that the geological epoch of the last 12,000 years (the Holocene) has given way to a new one bearing the stamp of *Homo sapiens*—the Anthropocene.[27] Furthermore, as Morton observes, global warming coincides with the Anthropocene as well as the sixth mass extinction, which is bringing an end to the Cenozoic era—the geological era that began approximately 65 million years ago following the extinction of the dinosaurs. "Maybe it would make it more obvious if we stopped calling it 'global warming' . . . and started calling it 'mass extinction,' which is the net effect."[28]

Involving all of those interrelated parts, climate change is massive. With global warming entering into social space, the very ground of being has shaken. A "*being-quake*" has happened because of global warming and other entities Morton calls "hyperobjects"—"things that are massively distributed in time and space relative to humans" (e.g., oil fields, the biosphere, *Homo sapiens*, global capitalism, the Internet, the Milky Way, and so on).[29] While hyperobjects are physically huge, Morton holds that a hyperobject is actually *smaller than the sum of its parts*. Of course, a whole is physically larger than its parts. A forest is larger than a tree. Earth's climate is bigger than a rainstorm. Wholes are big, but they exist in a way that disappears into their parts. Wholes might be spatially or temporally larger than their parts, but however "physically huge," they are "ontologically tiny."[30]

Try pointing at your hand, and you find yourself pointing at hand parts: fingers, fingernails, skin, palm, knuckles, and so on. Try pointing to a forest, and you find trees, soil, animals, and so on. Of course, a tree itself is a whole relative to its parts, and when you try pointing at a tree, you find yourself pointing at branches, bark, stems, leaves, roots, and so on. The whole is "literally out-scaled by its parts," and thus "however absurd and amazing it sounds, we need to say 'the whole is always smaller than the sum of its parts.'"[31] In contrast to holisms for which a whole transcends its parts and is greater than their sum, a whole for OOO withdraws and disappears: "it *subscends* its appearance in a way that is not constantly present."[32] Unlike the transcendence of "explosive holism," for which "the parts are reducible to the whole," subscendence involves an "implosive holism" in which every object can be seen as a whole and/or as a part of a larger whole, and no single whole matters inherently more than any other.[33] Every whole is vague and fragile—imperceptibly enmeshed with the parts on which it depends.

Biocentrism is right to criticize ecocentrism for occluding living organisms, but ecocentrism is right that whole ecosystems have value. The problem is that both biocentrism and ecocentrism think of ecosystems in terms of explosive holism, where the value of the whole transcends than the value of the parts. Ecocentrism affirms that holism, whereas biocentrism rejects its neglect of the living parts of an ecosystem. OOO thinks of wholes differently—implosively. Hospitality to strangers means welcoming disappearing wholes, including the forest ecosystem that disappears for the trees, the individual tree that disappears for the bark and branches, and global warming—the hyperobject that disappears for the weather, fossil fuel, fracking wells, rising sea levels, extreme weather events, scientific reports, economic policies, corporate lobbyists, and so on.

All wholes, at all scales, have equal value, and wholes do not transcend their parts but disappear into them. Moreover, saying that the climate disappears, subscends, or withdraws from weather does not mean that it is in some world behind the scenes, "empirically shrunken back or moving behind"; rather, it means that it is completely open, "*so in your face that you can't see it*."[34] A forest is so in my face, I can only point to trees. Global warming is so in my face, I can only point to extreme weather events, political debates, nongovernmental organizations, ocean acidification, and atmospheric concentrations of carbon dioxide and methane. Precisely this ungraspable intimacy is what makes it possible for ethical action to make a change. Every part of a hyperobject can play its part without being transcended or trumped by the whole.

Explosive holism disempowers objects from playing their parts, as can be seen in utilitarian, correlationalist, and Gaian holisms.[35] Utilitarianism is an explosive holism of the greater good, which is a sum of the goods of all sentient individuals, making it justifiable to kill one billion parts in order to save seven billion parts, to let a few island nations be submerged underneath rising seas in order to maintain the global economic order, or to implement carbon markets (e.g., cap and trade) that create economic value without actually mitigating climatic problems like deforestation and desertification. The correlationist holism is a Kantian one where the real is correlated to some subject, who becomes "the Decider" for all moral value, whether that value is centered in the autonomy of humans, the subjectivity of animal lives, or as biocentrism has it, the self-organization (autopoiesis) of organisms. Even at its most biocentric, the correlationist holism does not let inanimate entities play their parts. Biocentrism leaves crucial actors in the background, like weather, climate, carbon, methane, and ecosystems. Gaian holism is ecocentrism scaled up the whole planet, setting the transcendent bar for ethical value at the whole biosphere, making every entity a replaceable component of Gaia.

Implosive holism grants ethical consideration to every whole, at all scales, and it views every whole as a fragile entity imperceptibly enmeshed with its parts, such that my ethical concern for the climate is not separate from my concern for racial justice, gender equality, political elections, my personal health, the Pacific Ocean, future generations, poetry, and my cat. Become a vegan, advocate for environmental justice, pet your cat, or read a poem. You are always already engaging in actions that impact the climate, however imperceptible or statistically insignificant. Climate ethics requires attentive ways of engaging the interconnectedness of things (the mesh of strangers) without recourse to mega-centrisms or other categories of explosive holism, including ostensibly ecofriendly categories like "nature" or "matter." Morton (in)famously calls this kind of ecological thinking "ecology without nature," which is also "*ecology without matter*."[36] Global warming is crowding into social space, bringing humans unbearably close to the polymorphous plurality of things that ideas of "matter," "world," and "nature" used to keep in the background.[37]

Global warming is happening where I am, here and now, yet it is also happening over the course of centuries and across Earth's surface, radiating innumerable spatiotemporal ripples: political administrations, polar ice caps, research grants, a mass extinction event, seasons, information technologies, hurricanes, the hydrologic cycle, a conversation at my dinner table, and a drop of rain during a rainstorm. I feel a drop of water hit my head. That was not just a raindrop or even a rainstorm. "When it rains on my head, climate is raining," such that climate touches me indirectly.[38]

Earth's climate is a hyperobject that demands to be welcomed into human consideration, but it withdraws from direct contact and overwhelms the scales with which humans make ethical decisions. Indeed, it overwhelms the very boundaries of ethics. If all things are strangers, bearing ethically compelling otherness or weirdness, then all entities are engaging in ethics, forging hostile or hospitable relations with one another. When ethics affirms the irreducible plurality of beings, ethics beyond humans does not stop at animals, life, ecosystems, Gaia, or the universe. Ethics beyond humans is ultimately ethics without borders. As Ian Bogost puts it, "ethics itself is revealed to be a hyperobject: a massive, tangled chain of objects lampooning one another through weird relation, mistaking their own essences for that of the alien objects they encounter, exploding the very idea of ethics to infinity."[39] Ethics no longer has a center, not *anthropos*, *bios*, or *oikos*. Object-oriented ethics is ex-centric. This basically renders ethical decisions impossible, sending ethics into a vertiginous multiplicity of competing demands issuing across interconnected entities at all scales. The aporia of climate ethics is that hyperobjects like global warming issue imperatives to which humans must respond, yet there is no stable ground or orienting center from which to forge a response. In other words, "we have no time to learn fully about hyperobjects. But we have to handle them anyway."[40]

Spectral attunement

OOO does not ask you to decide on what you should focus ethical concern (e.g., humans, sentient beings, individual organisms, ecosystems, the biosphere). Everything has equal value, and everything is interconnected, so the question of climate ethics becomes less about *what* you do and more about *how* you do it. Responding to global warming requires a hospitable atmosphere. "Mood," "atmosphere," and "attunement" are all translations of the German noun *Stimmung*. The tuning of a guitar is its *Stimmung*, and the voice of a singer is her *Stimme*. For Immanuel Kant, aesthetic experience is described as a *Stimmung*.[41] For Heidegger's existential phenomenology, all experience involves *Stimmung*, such that mood is an ontologically primordial condition through which humans find themselves in the world. Like an existential affect or feeling, *Stimmung* is what "makes it possible first of all to direct oneself toward something."[42] Undoing the anthropocentrism of Kant and Heidegger, OOO finds attunement happening in all beings. Every object interacts through its own mood, its own tune or timbre.[43] When I listen to a song, I am tuning to it while it is tuning to me, searching out resonances within my body. Similarly, in the "sticky mesh of viscosity" that imperceptibly entangles humans with an object like climate change, "I find myself tuned by the object."[44] Smelling smoke, I am attuning to nearby fires in Northern California, and global warming is attuning to my nostrils.

Tuning is how all things interact, touching each other vicariously by resonating with each other's qualities; "attunement is the mode in which causality happens."[45] For Kant, beauty and the sublime are experienced in an accord that tunes the subject and object, harmonizing them and fusing them together. This fusion can be confusing and overwhelming to the subject, and insofar as this aesthetic dimension is extended to all objects, this fusion can lead to destruction and death. You can think of attunement like a virtue, which is a mean between two vices—a deficiency (out of tune; dissonance) and an excess (perfectly tuned; unison). If two things are too far out of tune, they become dissonant and disassociated, and if two things are too perfectly in tune, they become so closely associated that they crash into one another. Perfect attunement undoes the boundaries that keep things intact. "When an opera singer matches the resonant frequency of a glass, the glass explodes."[46] Resonating in perfect unison spells death for heterogeneity, otherness, strangeness, and uniqueness. Ethics is best when it's imperfect. Perfect ethics would be ruinous, hence Ludwig Wittgenstein's remark about how a totally realized book of ethics would resonate so perfectly with everything ever written on ethics that all ethics books would explode.[47]

Attunement is a tactful touch that makes contact with the other while letting the other remain intact. Object-oriented ethics cultivates a tender touch. It avoids any purity ethics that seeks a perfect answer or drives a wedge between a good *us* and ignoble or benighted *them*. OOO is in line with Alexis Shotwell, who takes a stand against the ethics and politics of "purism"—a "common approach for anyone who attempts to meet and control a complex situation that is fundamentally outside our control," which is to say, "it shuts down precisely the field of possibility that might allow us to take better collective action against the destruction of the world in all its strange, delightful, impure frolic."[48]

Impossible to handle perfectly, global warming has to be handled nonetheless. Preventing people from escaping that situation, global warming makes ethics hypocritical. "*The time of hyperobjects* is a *time of hypocrisy*."[49] The choice to care about global warming is hypocritical, since each person has a carbon footprint and is complicit to varying degrees in institutions and industries that also have carbon footprints. Moreover, the cynical who wants to opt out of this hypocrisy chooses to give up and avoid handling global warming at all, but that is still a choice. Global warming sticks to the cynic just as much as it does to everyone else. The cynic tries to avoid hypocrisy but ends up being hypocritical about hypocrisy.[50]

Cynical reason corresponds with a phase of ethical development that G. W. F. Hegel calls "the beautiful soul."[51] The beautiful soul is someone who seeks harmony while avoiding anything that would compromise or contaminate the purity of that ideal, which basically means avoiding everything in the external world. The beautiful soul is "entangled in the contradiction between its pure self and the necessity of that self to externalize itself and change itself into an *actual* existence."[52] "Beautiful soul syndrome," as Morton terms it, names the position of the cynic in the Anthropocene. "Beautiful me over here, corrupt world over there."[53] Much to the beautiful soul's chagrin, that view *is* the corruption it sees in the world. In other words, supposedly "bad" people may be a driver of collective inaction regarding global warming, but an even bigger driver may be the view that draws a rigid line separating people into "good" or "bad."

Whereas the beautiful soul seeks unambiguous and harmonious decisions, one who accepts the hypocritical conditions of decisions seeks a stranger mood, an attunement more fit for specters, a "floating of decision" in "spectral attunement."[54] Along with hypocrisy, this spectral attunement is characterized by weakness and lameness, "*weakness* from the gap between phenomenon and thing," and "*lameness* from the fact that all entities are fragile."[55] The gap between a thing and its appearance is the space of attunement, which involves weakness insofar as nothing can make direct contact with anything. I cannot do anything directly to Earth's climate, and yet everything I do indirectly impacts it. The fragility of things is their subscendence, whereby things are inconsistent, torn into pieces, failing to coincide with themselves, in the way that rain is and is not climate, and I am and am not *Homo sapiens*.[56]

The weakness, lameness, and hypocrisy of spectral attunement means everything I do about global warming is inadequate and even self-defeating. Creating more bike lanes in a dense city like San Francisco can motivate more people to ride a bike, thus reducing carbon emissions from personal automobiles, but if the construction of the bike lane involves eliminating public parking spaces, it could cause an overall increase in emissions while more people are circling and idling while looking for parking. When everything is interconnected, unintended consequences are everywhere. That is partly why Morton calls his ecological thought *dark* ecology. Nothing works. That darkness is depressing, but fighting the depression only causes more problems, reacting against one's own moods and against the conditions of coexistence in the Anthropocene. "Don't fight it," advises Morton, but try to "tunnel down" so that the darkness becomes more mysterious than depressing and finally becomes "dark and sweet like chocolate."[57] Morton discusses a series of tunings that move through dark ecology, beginning with guilt and shame.[58] People are stuck inside global warming. For those who are cynical about that complicity, feelings of guilt and shame arise. Accepting that complicity is depressing, leaving you feeling the unbearable imprints of all the beings enmeshed in this massive problem.

The more you accept the melancholy of this situation, the more horrifying it becomes. The horror starts to appear ridiculous after a while, as it becomes obvious that your attempt to find an anthropocentric escape from ecological horror *is* ecological horror. The attempt of humans to disengage from the ecological mesh is "the Severing."[59] It produces the anthropocentric attitudes that began operating in civilization in the Neolithic development of agriculture and were subsequently exacerbated by the modern risk society and global capitalism. The Severing causes ecological problems to which humans respond by trying to sever their connections to ecological problems, thus producing more ecological problems and more severing, and so on unto mass extinction.

When you realize that you are the ecological loop you are trying to escape, then tragedy starts feeling like comedy. In the space of comedy, the wall separating humans from the mesh is shown to be ridiculous, and the overwhelming horror of global warming now appears as a more ambiguous or absurd strangeness. The laughter that comes with ridicule can give way to interest and fascination. "Fascinated, I begin to laugh with non-humans, rather than at them (horror and ridicule), or at and with my fellow humans about them (shame and guilt)."[60] That fascination opens onto a deeper sadness, the sadness of beauty. It is the sadness of attuning to that which you can never grasp. Within that sadness is a longing, which subtends the "*basic anxiety*" manifest in guilt, shame, and horror, and ridicule.[61]

"We have anxiety because we care."[62] The withdrawal of things is what makes it possible for humans to care about them, but it also renders care impossible. Care is a longing for the stranger, one who is always coming but never shows up. Caring about global warming makes you anxious, and when you try deleting that anxiety so you care perfectly right or not at all, your attunement will be something like guilt, shame, sadness, horror, or ridicule, but if you accept that anxiety, the ongoing frustration of longing can feel playful and fun. Care becomes a little careless or carefree. It is a *"playful care,"* "care with the care/less halo," a spectral care that indicates not a lack of seriousness but a *"playful seriousness,"* open to the ambiguous play whereby global warming appears and withdraws in all aspects of human existence.[63] Can climate ethics become something you enjoy? Playful care might work better than guilt and shame when communicating with people who deny the reality of anthropogenic climate change, and it might lighten the tremendous burden borne by those who already care about the climate. Imagine someone petting a cat, bonding with the cat for the sheer joy of it. That person is cultivating a nonviolent style of attuning to non-humans. That is not a serious response to global warming, and yet it is not ineffective. When someone learns to enjoy nonviolent coexistence with non-humans, that mood mediates that person's practices and habits. If all humans thoroughly enjoyed sharing time with non-human neighbors, institutions and industries would reflect that mood, just like today's institutions and industries reflect a predominantly anthropocentric mood.

To be sure, joy still involves pain, grief, mourning, and unrequited longing, hence the darkness of this joy. When caring becomes joy, the darkness of dark ecology tastes bittersweet, like dark chocolate. Where guilt, shame, and depression narrow one's capacity to act, joy can energize ongoing engagements in the tediously local yet massively international and intergenerational task of mitigating and adapting to Earth's changing climate. Where explosive holisms assimilate the plurality of objects, joy is a mood that lets itself be tuned by strangers. Insofar as non-humans are always already tuning you, joy is always already happening, whether you pay attention or not. It is a "basic effervescence" accompanying the basic anxiety of coexistence.[64] Joy does not get rid of that anxiety. It is what anxiety feels like when you let it well up and do not try to erase it. It feels like a hospitable atmosphere, like a strange solidarity with non-humans, where "solidarity is the default affective environment of the top layers of Earth's crust."[65] With this solidarity, Morton advocates for an ecological communism or "ecocommunism, a communism of humans and non-humans alike."[66] Solidarity is the default atmosphere of the symbiotic real. Since you are always already symbiotic, finding solidarity is a challenge not because it is too difficult but because it is too easy.[67]

How can humans tune into that solidarity and find joy in the struggle for nonviolent coexistence? One area of human experience excels in inspiring joy amidst the anxiety of never-ending longing: consumerism. It gets boring to chase after new products and trends all the time, always seeking the next new thing. Yet, boredom with the treadmill of consumer pleasure is part of what people enjoy. This ennui is the main ingredient in consumerist experience, "stimulated by the boredom of being constantly stimulated."[68] Ennui is the atmosphere with which humans can attune to Earth's atmosphere. "Ennui is the correct ecological attunement!"[69]

Consumerist experience and the capitalist economy of which it is a part are among the main forces driving the climate crisis. Consumerism seems antithetical to climate ethics, especially in light of the beautiful soul syndrome of scholars and activists who are disgusted by consumerism, but the ennui consumerism inspires is strangely the right mood through which humans can find joy and solidarity in their responses to global warming. Boredom does not try to erase the basic anxiety of not knowing how to interact with objects. The problem with capitalism is not its endless loop of longing. The problem is that *"capitalism is not spectral enough."*[70] Consumerism does not seek enough enjoyment. Consumers are entranced by specters that take on lives of their own, like apps, clothes, cars, shoes, iPhones, and so many other products. Consumers are letting non-humans have agency and even have control over human lives, and humans are somewhat bored and somewhat pleased by it. That simply needs to be opened up to more non-humans. There are countless more specters with which one can interact, like rivers, chimpanzees, the biosphere, the atmosphere, neighborhoods, compost heaps, and electrical grids. The slightly disgusted enjoyment of a bored consumer is careless and carefree, unburdened by cries that seek relief from anxiety, like the classic ethical cry, "What are we going to do?"[71] Intimacy with non-humans has less to do with asking questions to figure out what should be done and more to do with listlessness, becoming highly susceptible to non-humans.

Listlessness is not normally considered a good thing in ethics. Being listless bears connotations of being unmotivated, ineffectual, uncaring, and lacking energy or enthusiasm. Listlessness seems counterproductive to living a good life, and it seems catastrophic if it is applied to global warming. However, listlessness becomes a virtue once it becomes clear that efforts to get out of ecological crisis (the Severing) perpetuate systems that produce ecological crisis. It is like a Chinese finger trap, a puzzle that you play with by putting a finger from one hand in one side of a small, finger-sized tube, and putting a finger from the other hand in the other side. Once your fingers are inside, you cannot pull them out without the trap tightening around your fingers and thus further entrenching you in the trap. The only way out is through, to let go, to go with the flow. If you let your fingers move further into the trap, the trap relaxes its grip and you can effortlessly liberate your fingers. Freedom comes from accepting the trap.

Sincere ecological attunement is ironic. It is ironic that the boredom characterizing consumerist experience, which is driving global warming, also welcomes the irreducible strangeness of things and thus makes possible non-anthropocentric ethical responses to global warming. It is ironic that trying to care gets in the way of actually caring, and that some carelessness is required for welcoming beings on their own terms. This attunement is about "realizing the irony of being caught in a loop and how that irony does not bestow escape velocity from the loop. Irony and sincerity intertwine. This irony is joy, and the joy is *erotic*."[72]

Joy is the polymorphous perversity of parts exceeding their wholes. It is the erotic dance whereby things appear and disappear, flickering with deep pasts and uncertain futures. Coexistence is drenched in pleasure. "Ecology is *all* delicious," all "*ecosexual*," including the guilt, shame, melancholy, horror, sadness, longing, and joy.[73] An object-oriented approach to climate ethics is about multiplying enjoyments, inventing new pleasures for coexistence. You can feel more connected with Nature, become more efficient with your use of resources, make art that raises awareness about global warming, and behave as a role model for others.[74] However, if the practices of connectedness, efficiency, art, or role modeling feel compulsory, then find something else to do. It is less about what you do than how you do it. Whatever you do, consider your style.

Object-oriented climate ethics is not asking everyone to be an activist or asking everyone to care about climate change or environmental issues. It is not asking everyone to be friends and companions. It is simply asking people to lighten the mood, to accept coexistence on a warming planet amidst a polymorphous plurality of things; to accept the ambiguity, irony, weirdness, and strangeness of things; to accept the guilt, shame, depression, horror, and joy that come with caring; and to accept the state of acceptance, which can feel boring, careless, and listless. Such an approach to climate ethics will never solve global warming, but it may open up different ways of letting this overwhelmingly massive object into people's lives. It may open up ways of relating to Earth's climate as something other than a problem to solve, something stranger than a crisis from which to escape, something you might even enjoy learning to live with.

Notes

1 Brian Swimme and Mary Eveyn Tucker, *Journey of the Universe* (New Haven: Yale University Press, 2011), 17.
2 Jeff Bridges, *Living in the Future's Past*, dir. Susan Kucera (Marina Del Rey: Vision Films, 2018).
3 Morton calls anthropogenic climate change "global warming," as it is not just a generic change but specifically a warming trend. Timothy Morton, *Hyperobjects: Philosophy and Ecology After the End of the World* (Minneapolis: University of Minnesota Press, 2013), 3.
4 Timothy Morton, *Dark Ecology: Toward a Logic of Future Coexistence* (New York: Columbia University Press, 2016), 131.
5 Timothy Morton, *The Ecological Thought* (Cambridge: Harvard University Press, 2010), 15.
6 Edith Wyschogrod, *Emmanuel Levinas: The Problem of Ethical Metaphysics*, 2nd ed. (New York: Fordham University Press, 2000).
7 Timothy Morton, *Being Ecological* (Cambridge: MIT Press, 2018), 70.
8 Timothy Morton, "Deconstruction and/as Ecology," in *The Oxford Handbook of Ecocriticism*, ed. Greg Garrard (Oxford: Oxford University Press, 2014). For more ecological interpretations of Derrida's deconstruction, see Matthias Fritsch, Philippe Lynes, and David Wood, eds., *Eco-Deconstruction: Derrida and Environmental Philosophy* (New York: Fordham University Press, 2018).
9 Morton, *Being Ecological*, 67.
10 Morton, *The Ecological Thought*, 143–72. Jacques Derrida, *Specters of Marx: The State of the Debt, the Work of Mourning, and the New International*, trans. Peggy Kamuf (New York: Routledge, 1994), 196.
11 Derrida, *Specters of Marx*, 10, 161.

12 Martin Heidegger, *Being and Time*, trans. John Macquarrie and Edward Robinson (New York: Harper & Row, 1962), 99.

13 Morton, *Being Ecological*, 7.

14 A key difference between Harman and Morton is that the former draws little from Derrida and is highly critical of what he takes to be Derrida's anti-realism. Among the various proponents of OOO, Morton is the one who is most indebted to Derrida. Graham Harman, *Object-Oriented Ontology: A New Theory of Everything* (London: Pelican, 2018), 199.

15 Ibid., 80, 150–52.

16 Ibid., 41–52.

17 Ibid., 150.

18 Timothy Morton, *Realist Magic: Objects, Ontology, Causality* (Ann Arbor: Open Humanities Press, 2013), 20.

19 Harman, *Object-Oriented Ontology*, 54.

20 Morton, *Being Ecological*, 126–30. Cf. Timothy Morton, *Humankind: Solidarity with Nonhuman People* (London: Verso, 2017), 135.

21 Morton, *Being Ecological*, 128.

22 Jacques Derrida, "Hostipitality," in *Acts of Religion*, ed. Gil Anidjar (New York: Routledge, 2002), 364. Derrida derives his sense of hospitality from Emmanuel Levinas, for whom hospitality is "the one-for-the-other in the ego," which means "giving to the other the bread from one's own mouth" and "being able to give up one's soul for another." Emmanuel Levinas, *Otherwise than Being, or, Beyond Essence*, trans. Alphonso Lingis (Pittsburg: Duquesne University Press, 1998), 79.

23 Derrida, "Hostipitality," 419.

24 Anthony Weston, "Multicentrism: A Manifesto," *Environmental Ethics* 26, no.1 (2004): 36.

25 Ibid., 35.

26 Ibid.

27 As the climate scientist Mike Hulme puts it, climate change as a planetary system of interconnected weather patterns is a part of a more complex whole that includes 1) what Ulrich Beck analyzes as the modern "risk society," 2) capitalist ideology, 3) the end of nature, and 4) the Anthropocene. Mike Hulme, "(Still) Disagreeing About Climate Change: Which Way Forward?" *Zygon* 50, no. 4 (2015): 897–99.

28 Morton, *Being Ecological*, 5.

29 Morton, *Hyperobjects*, 1, 19. Determining where objects stop and hyperobjects begin can be difficult. Any object could appear hyper in relation to another of a different scale. To some extent, "every object is a hyperobject." Ibid., 201.

30 Morton, *Being Ecological*, 102.

31 Morton, *Humankind*, 106.

32 Ibid., 106.

33 Ibid., 70.

34 Ibid., 37.

35 Ibid., 121–22.

36 Morton, *Hyperobjects*, 92. Moreover, Morton is not aiming for total transcendence of these terms, and he is not trying to take away anybody's favorite words. "I balk at saying *without* in the sense of 'utterly without' or 'beyond'"; a "formula like that tries to progress once and for all like the modernity of which it is sick." Morton, *Dark Ecology*, 83.

37 Morton, *Humankind*, 122.

38 Morton, *Hyperobjects*, 76.

39 Ian Bogost, *Alien Phenomenology, or What It's Like to Be a Thing* (Minneapolis: University of Minnesota Press, 2012), 79.

40 Morton, *Hyperobjects*, 67.

41 Immanuel Kant, *Critique of Judgment*, trans. Werner S. Pluhar (Indianapolis: Hackett, 1987), 445–46.

42 Heidegger, *Being and Time*, 176.

43 Morton, *Being Ecological*, 94.
44 Morton, *Hyperobjects*, 30.
45 Morton, *Being Ecological*, 90.
46 Morton, *Realist Magic*, 193.
47 In Wittgenstein's words: "if a man could write a book on Ethics which really was a book on Ethics, this book would, with an explosion, destroy all the other books in the world." Ludwig Wittgenstein, "A Lecture on Ethics," *The Philosophical Review* 74, no. 1 (1965): 7.
48 Alexis Shotwell, *Against Purity: Living Ethically in Compromised Times* (Minneapolis: University of Minnesota, 2016), 7–9.
49 Morton, *Hyperobjects*, 6.
50 Ibid., 148.
51 Georg Wilhelm Friedrich Hegel, *Phenomenology of Spirit*, trans. A. V. Miller (Oxford: Oxford University Press, 1977), 406.
52 Ibid.
53 Morton, *Hyperobjects*, 154.
54 Morton, *Humankind*, 82.
55 Morton, *Hyperobjects*, 2.
56 Ibid., 195.
57 Morton, *Dark Ecology*, 117.
58 Ibid., 131–35.
59 Morton, *Humankind*, 13–18.
60 Morton, *Dark Ecology*, 147.
61 Morton relates this notion of basic anxiety, coined by the Buddhist teacher Chögyam Trungpa, to Heidegger's notion of anxiety (*Angst*) as an attunement that discloses the fundamental openness of human existence. Morton, *Realist Magic*, 203.
62 Morton, *Dark Ecology*, 152.
63 Morton, *Being Ecological*, 131.
64 Morton, *Dark Ecology*, 155.
65 Morton, *Humankind*, 14.
66 Ibid., 55.
67 Morton, *Being Ecological*, 157.
68 Morton, *Humankind*, 65.
69 Ibid., 66.
70 Ibid., 63.
71 Morton, *Being Ecological*, xxvii.
72 Morton, *Dark Ecology*, 155.
73 Ibid., 129.
74 Those attunements or "styles" are what Morton calls, respectively, "immersive," "authentic," "religious," and "efficient." Morton, *Being Ecological*, 139–52.

References

Attfield, Robin. "Biocentrism, Climate Change, and the Spatial and Temporal Scope of Ethics." In *Climate Change Ethics and the Non-Human World*, edited by Brian G. Henning and Zack Walsh (this volume).

Bogost, Ian. *Alien Phenomenology, or What It's Like to Be a Thing*. Minneapolis: University of Minnesota Press, 2012.

Bridges, Jeff. *Living in the Future's Past*. Directed by Susan Kucera. Marina Del Rey: Vision Films, 2018.

Derrida, Jacques. "Hostipitality." In *Acts of Religion*, edited by Gil Anidjar, 356–420. New York: Routledge, 2002.

Derrida, Jacques. *Specters of Marx: The State of the Debt, the Work of Mourning, and the New International*. Translated by Peggy Kamuf. New York: Routledge, 1994.

Fritsch, Matthias, Philippe Lynes, and David Wood, eds. *Eco-Deconstruction: Derrida and Environmental Philosophy*. New York: Fordham University Press, 2018.

Harman, Graham. *Object-Oriented Ontology: A New Theory of Everything*. London: Pelican, 2018.

Hegel, Georg Wilhelm Friedrich. *Phenomenology of Spirit*. Translated by A.V. Miller. Oxford: Oxford University Press, 1977.

Heidegger, Martin. *Being and Time*. Translated by John Macquarrie and Edward Robinson. New York: Harper & Row, 1962.

Hulme, Mike. "(Still) Disagreeing About Climate Change: Which Way Forward?" *Zygon* 50, no. 4 (2015): 893–905.

Kant, Immanuel. *Critique of Judgment*. Translated by Werner S. Pluhar. Indianapolis: Hackett, 1987.

Levinas, Emmanuel. *Otherwise than Being, or, Beyond Essence*. Translated by Alphonso Lingis. Pittsburg: Duquesne University Press, 1998.

Morton, Timothy. *Being Ecological*. Cambridge: MIT Press, 2018.

Morton, Timothy. *Dark Ecology: Toward a Logic of Future Coexistence*. New York: Columbia University Press, 2016.

Morton, Timothy. "Deconstruction and/as Ecology." In *The Oxford Handbook of Ecocriticism*, edited by Greg Garrard, 291–304. Oxford: Oxford University Press, 2014.

Morton, Timothy. *The Ecological Thought*. Cambridge: Harvard University Press, 2010.

Morton, Timothy. *Ecology Without Nature: Rethinking Environmental Aesthetics*. Cambridge: Harvard University Press, 2007.

Morton, Timothy. *Humankind: Solidarity with Nonhuman People*. London: Verso, 2017.

Morton, Timothy. *Hyperobjects: Philosophy and Ecology After the End of the World*. Minneapolis: University of Minnesota Press, 2013.

Morton, Timothy. *Realist Magic: Objects, Ontology, Causality*. Ann Arbor: Open Humanities Press, 2013.

Nichols, Amanda. "Being Human: An Ecocentric Approach to Climate Ethics." In *Climate Change Ethics and the Non-Human World*, edited by Brian Henning and Zack Walsh (this volume).

Rolston, Holmes. "Wonderland Earth in the Anthropocene Epoch." In *Climate Change Ethics and the Non-Human World*, edited by Brian Henning and Zack Walsh (this volume).

Shotwell, Alexis. *Against Purity: Living Ethically in Compromised Times*. Minneapolis: University of Minnesota, 2016.

Swimme, Brian and Mary Eveyn Tucker. *Journey of the Universe*. New Haven: Yale University Press, 2011.

Weston, Anthony. "Multicentrism: A Manifesto." *Environmental Ethics* 26, no.1 (2004): 25–40.

Wittgenstein, Ludwig. "A Lecture on Ethics." *The Philosophical Review* 74, no. 1 (1965): 3–12.

Wyschogrod, Edith. *Emmanuel Levinas: The Problem of Ethical Metaphysics*. 2nd ed. New York: Fordham University Press, 2000.

12 Monsters, metamorphoses, and the horror of ethics in the "Pelagioscene"

Jeremy Gordon

Living on land we sometimes forget the sea's dominance of our physical and cultural histories. We should remember.

— Steve Mentz, *At the Bottom of Shakespeare's Ocean*

Since childhood, I've been faithful to monsters. I have been absolved by them, because monsters, I believe, are patron saints of our blissful imperfection.

— Guillermo del Toro

Figure 12.1 Mermaid at the Edge

On the shore

I took this picture while at a conference in Copenhagen. Visiting the statue was an afterthought, a last-minute attempt to get a visual souvenir for my niece, a fan of *The Little Mermaid*. Revisiting the photo now, the chilly December visit was more than that. In fact, it was most fitting that I encountered the creature at that particular moment. I was attending the "Aesthetics, Ethics, and Biopolitics of the Posthuman" conference at Aarhus University, an inter-disciplinary gathering of international scholars engaged with how philosophical and artistic renderings of posthumanism shape and reshape political/ethical relations between human and more-than-human life worlds.[1] For three days, we attempted to synthesize and nuance "posthumanism," especially as it relates to emerging ethical orientations that de-center anthropocentric approaches to self, other, and ecology. Carey Wolfe's conception of posthumanism was central, especially his notion that posthumanism is not a disavowal of "the human" but a critical rejection of anthropocentric divisions between human, animal, vegetable, mineral, and technological.[2] For Wolfe, the human is simply another form of life among others. This kind of critical philosophical posthumanism, as Amanda Nichols summarizes in this edited volume, "expands common understandings of moral inclusion and extends rights to all other species. It looks beyond the human to reconceive notions of intelligence, consciousness, cooperation, sentience, and emotion."[3] According to Nichols, Eduardo Kohn and Donna Haraway foster posthuman ecocentric orientations, as each amplify the ways in which human and more-than-human bodies mix, mingle, and make complex webs of relationality and becoming, expanding who and what is worth moral consideration. In other words, if "we" are entangled with all kinds of biologies and bodies, it becomes difficult to establish a limit for who and what warrants ethical concern. Citing Val Plumwood, ecocentric posthumanism troubles categorical dualisms—nature/culture, human/animal, male/female, self/other, body/mind, et cetera—that help shape the "mastery model" of environments, in which an ideal, atomistic subject is apart from, and in control of, "mere" matter.[4]

The sea creature, even if bronzed and protected against the elements, troubles (some of) those dualisms. Between fish and human, water and rock, the mermaid might be encountered as an embodiment of the kind of posthuman becoming that Rosi Braidotti describes.[5] According to Braidotti, we live metamorphically. Metamorphosis *is* the condition of posthuman life. Corporeal modifications, altered genes, flows of capital, fluctuations in toxicity, troubled gender binaries, and unstable soil only stress the fact that divisions between water, rock, fish, and female quickly dissolve—especially in watery settings, where microplastics, industrial toxins, and other imperceptible matter swirl and seep through porous skin.[6] Recalling the little mermaid's tale/tail, morphing between human and sea creature is central, as is a challenge to the mastery model of pelagic ecologies. Encountering the little mermaid at the edge of a rocky water line was an encounter with the "strange stranger," what Sam Mickey might understand as a haunting figure that entangles us in an un-easy ethical relationship with alterity.[7] More than that, writing as part of this edited collection, Mickey argues that the presence of non-human entities situates "human ethics within a massive mesh of ethical injunctions of hospitality to strangers," especially those who attune us towards ecological attachment and the limits of atmospheres created by anthropocentric (capitalist) orientations.[8] With an object-oriented ontology (OOO) attitude, ethics becomes a process of altering the affective relations, the atmospheres, of coexistence. Joy, disgust, and boredom "with the kaleidoscopic scales of strange strangers coexisting in a warming climate" is how Mickey puts it.[9]

At issue is finding figures and stories for altering affective attunements and the atmospheres they help create. Highlighting the double-meaning of "scales," I argue that looking to the fins of sea creatures helps shape the un-easy ecological ethics that Mickey calls for, particularly as this essay is inspired by Haraway's "SF" model of ecological relations. In Haraway's terms, speculative fiction and feminism hold potential for imagining how to *become with* strange ecological others in ways that challenge anthropocentric ethics and the narrow categories of regard that might result.[10] Following Haraway into watery realms, we encounter strange-strangers—sea creatures that take us well beyond narrow versions of life, love, and land-based ethics. Watery bodies in speculative fiction, as we will see, attune us to the joys and pleasures of strange encounters. To be certain, pleasure is vital for animating care and regard for aquatic spaces and the scaly, slimy bodies that may provoke disgust and seemingly incommensurable difference. As such, Stacy Alaimo and Astrida Neimanis also accompany this quick dive into an aquatic approach to non-human climate ethics. Haraway, Alaimo, and Neimanis share a deep concern for how marine ecologies inform who "we" are, who we might become, and how radical exposure to watery bodies remind us that "we are all bodies of water," dynamic and fluid, porous and always already enmeshed in aquatic flow.[11] In the words of Alaimo, an "aqueous posthumanism . . . challenges us to imagine how the 'human,' at the level of the gene, sloshes around with the rest of oceanic life."[12] We are exposed. And to practice "insurgent vulnerability is to perform material rather than abstract alliances, and to inhabit a fraught sense of political agency that emerges" with a perceived loss of protected subjectivity.[13] For Alaimo, a politics of exposure troubles a Western model of disembodied mastery that grants detachment and solid grounding. More than this, however, performing exposure can animate "pleasure, desire, sensuality, and eroticism" as "permeating environmentalist ethics and politics."[14] This chapter, then, wades into how a pleasurable aqueous ethics of exposure might emerge, with the help of "monsters."

Wading in

Looking again, I see my photograph from Denmark displays distance from watery bodies. Even as I was exposed to coastal winds, standing on shore and looking down suggests being invulnerable from the waves that wash sea creatures ashore, safe from immersion, and certainly not close enough to notice the shimmering scales and the curl of fins. Still, the call of the strange siren atop the rock enchanted onlookers. Even as a much more majestic statue of Poseidon towered over onlookers just down the path, onlookers took more precarious paths just to get closer to the little mermaid. Slipping over rocks and risking a plunge into icy waters, we worked down the embankment and strained for the perfect picture, or a small touch of scales. The power of the fabled, stone-cold siren invites me to consider how sea creatures become guides for meaning making, for how water and oceanic realms come to matter, and thus, the potential such monsters have for enchanting us into more-than-human water ethics. My encounter with the mermaid in Denmark ushered me into more watery worlds, especially the aquatic shows of old Florida at Weeki Wachee Springs.[15] There are mermaids at Weeki Wachee. You can watch them perform underwater twice a day. Becoming a mermaid at Weeki Watchee is respected. Performances are part of a storied tradition.[16] Performers maintain a significant level of notoriety, which has been leveraged to promote environmental concern. More specifically, Weeki Wachee mermaids have become entangled with manatee protection efforts. Photos of mermaids swimming with manatees, mixed with statements and stories, help image how fiction and reality swirl together for shaping ethical concern. Fantastic sea creatures help create affinities with very real manatees—and the Florida springs in which they drift.[17] The manatee/mermaid relationship should not be all that surprising. Sailors used to see manatees as mermaids. When manatees surface, the silhouette looks remarkably human. While we might now see such sights as a "mistake" of vision or a trick of the eyes, at Weeki Wachee, we are encouraged to be mistaken, to see speculatively: "If you thought mermaids were just the lively imaginings of lonely sailors, think again – and come to Weeki Wachee Springs, the City of Live Mermaids, on the Gulf Coast of Florida." At Weeki Wachee, manatees and mermaids are part of a fantastic cast of swimmers that promotes seeing oceans as strange and full of life that is worth attention, care, and protection.

Mermaids, in Denmark and Florida, emerge as harbingers of category crisis, forms that evade enclosures, bifurcations, and taxonomical systems, forms that call into question anthropocentric assumptions of knowledge that can justify objectification and instrumentalization.[18] Mermaid tales/tails in Denmark and Florida animate deep devotions to sea-monsters and the kinds of oceanic encounters they (might) create. Strange, more-than-human beings animate questions of how seas come to matter, in both senses of the term: how pelagic realms matter to people and how oceanic matter emerges; how social imaginaries are shaped and how social imaginaries influence treatments of the ocean. More specifically, my concern is how sea-monsters—fantastic and not—immerse us with/in deeper pelagic orientations. This chapter drifts in those directions. Or rather, it is lured to follow the soaked sights and sounds that might offer a finned and web-footed way towards non-human aquatic ethics, fitting for life in what we might call the *pelagiocene*, the oceanic side of what has been called the Anthropocene. Rising tides of climate change, plastic-filled whales, and underwater extraction (allied with military sonic blasts) signal sea changes wrought by human practices of capital flow. Human systems have altered Earth's geology, as well as its hydrology. While the Anthropocene is commonly grounded in terrestrial terms, emphasizing the fluid mechanics of the epoch helps us fathom a posthuman scene of climate change.[19] Water resists, transgresses, floods, and flows over human geographies, stressing ways we stay afloat. Stacy Alaimo argues that descending into deep waters—the bathypelagic, abyssopelagic, and hadal zones of oceans—challenges anthropocentric epistemologies and the categories used to determine "life."[20] Encountering radical oceanic otherness can be pleasurable, stresses Alaimo, if we remain radically exposed. How are such pleasurable moments possible? Where do we find such encounters? What sorts of exposures "would foster posthuman anti-consumerist subjectivities" and the intimacies necessary for more ethical sensitivities?[21]

Sinking to encounter sea-creatures in the "low" culture of popular cinema provides one sort of exposure to posthuman ethics. Thus, this chapter highlights how pop culture sea-monsters provide glimpses at the violence of what Donna Haraway calls the "Capitalocene," the geologic age shaped by accelerated, expansive, and unequal extraction, production, and consumption of ecologies.[22] In particular, I focus on how sea-creature films prompt un-expected encounters with creatures that feature the possibility of monstrous unions with watery bodies. In horror films—like *The Creature from the Black Lagoon* and *Swamp Thing*, among others—cross-ontological alliances and strange couplings become central, especially those that explicitly and erotically challenge mastery of aquatic spaces. According to Eric White, such posthuman films affirm the "prospect of becoming" more monstrous as something to be desired.[23] Exploring the seascapes of popular posthuman cinema helps imagine more entangled, intimate, and erotic watery relations, making for a radically non-human aquatic ecology teeming with "alien" life worthy of consideration. Low theory submerges us in the deep, where we might find concern for creatures radically strange but certainly not inferior.

Diving, riding waves, submerging, and surfacing, sea creatures in popular culture emerge to help us imagine vibrant dimensions of existence that might otherwise be dismissed as a lifeless abyss open for plunder and capitalist waste.[24] The social construction of the ocean as a "vast void" and an empty surface for transportation has helped turn the sea into a horizontal space for questing, conquering, and commerce.[25] In response, we seek a more vertiginous orientation, sensing diverse worlds might exist and flow beneath ocean surfaces.[26] Fictional monsters from the deep help show us that the "human form is simply one composition among many, not the measure of the world."[27] For Haraway, speculative fiction provides one way to see the vibrant sympoetic crossings and multi-species assemblages that make up our ever-emergent ecological selves.[28] Speculative fiction, in the form of posthuman cinema, features entangled ontologies, situating "human" bodies in "the mesh," as Timothy Morton describes it.[29] In the mesh of posthuman film, permeability rules.[30] Constant flux between inside and outside center bodies as inescapable ecological.[31] Ecological bodies are especially permeable in watery worlds, where bodies in/of water seep into each other.[32] Most important, posthuman cinema affirms enmeshment. In speculative monster films, to cite Mickey, "coexistence is drenched with pleasure."[33]

Steve Mentz argues that living with/in ocean-drenched environments requires altering our fictions, replacing narratives of control to "less epic, more improvisational stories. . . . We need sailors and swimmers to supplement our oversupply of warriors and emperors."[34] While I agree with Mentz, sailors and swimmers are still a bit human for my taste, particularly when sailors and swimmers are deployed for empire. I am not alone. Haraway proposes more monstrous figures. She desires speculative fictions featuring mythic gorgons that destroy "the twenty-first century ships of the heroes on a living coral reef instead of allowing them, to suck the last drop of fossil fresh out of dead rock."[35] Haraway finds potential in the medusan, chthonic ones—tentacular creatures that challenge skyward-looking, surface-loving Anthropos. Connections between gorgons and seas take us into what Haraway calls the Chthulucene, the age of tentancled creatures and monsters that emerge on the heels of capitalism. Here, gorgons, octopuses, vampire squids, and jellyfish guide us into underworlds and underwaters, from which messy, unfinished eruptions of life drown out narratives of progress and innovation: "They are good figures for the luring, beckoning, gorgeous, finite, dangerous precarities" of contemporary ecology.[36] In the Chthulucene, human life is entangled with the "sheer not-us"—sea witches, fishy creatures, and ocean goddesses that demand attention.[37] Monsters "lure us into less anthropocentric, less 'grounded' modes of knowledge, politics, and ethics."[38] They force us to confront anthropocentric limits of thought, speech, and life, immersing us in a "swirled mess of obligation," where we might feel out of our depth.[39]

Here, then, there be monsters. Perhaps the very monsters necessary for a non-human oceanic climate change ethics. In this essay, I argue that we embrace the "horror" of climate change, situating ethics as constituted by creatures that emerge from the deep. Specifically, I turn to monsters of what Judith Halberstam calls the "low theory" of popular culture, where sea monsters inspire more dynamic ways of relating to oceanic life worlds.[40] The remainder of the chapter explores how a piece of posthuman popular culture attempts to cultivate deeper relations with bodies of water. Going overboard with mermaids leads us into the queer ecologies of Guillermo del Torro's *The Shape of Water*, where we find pleasure in strange, saturated affinities that contest bio-technical-militarism, mastery, and anthropocentric violence. As a story of radical hospitality of strange strangers and the pleasures of aqueous exposure, *The Shape of Water* teaches us how to live and love in the pelagiocene.

The shape of water: fathoming creaturely depths

> The most intelligent of creatures often make the fewest sounds.
> – Guillermo del Toro, *The Shape of Water*

Guillermo del Toro's critically acclaimed 2017 film *The Shape of Water* ends at the edge of a harbor, where a sea-creature is attempting to escape into a blue-black sea. The creature cannot survive outside the water for long. As we learn from laboratory encounters earlier in the film, it is able to breathe air but ultimately requires a very specific blend of salt water to flourish. The monster has hints of what might be classified as "human"—legs and arms and upright posture. However, it also has webbed extremities, gills, scales, teeth that tear flesh, and eyes that open and close in un-expected ways. The creature is joined by a woman, Elisa, who has helped the creature to this point. The scene is somewhat ominous. It is a dark night, and rain pours over concrete, making it difficult to discern where the ocean begins and the harbor ends. Elisa and the creature are drenched, soaking wet, vulnerable, and exposed—to the elements and to security personnel on their way to re-capture "the asset." Their goodbye is longer than it should be, providing us with a glimpse of what Stacy Alaimo describes as immersed subjectivity, in which bodies and minds are mired, intermeshed in circulating political and ecological currents.[41] For a drenched moment, we sense an intense attachment with a sea monster, attraction to a "thing" from the deep, intimacy that denies anthropocentric visions of relationality. The soaked scene stresses that, in the words of Astrida Neimanis, "our organs remember, that our bodies are archives" of watery relations.[42] "Water will exceed us – our sovereign . . . bodies, and our capacity to fully know."[43] Water does exceed us in *The Shape of Water*. It exceeds everyone and everything.

"Soaked" is one way to describe del Toro's film. From beginning to end, water is pervasive, creating a liquid optic through which we watch, and feel, an erotic-ethical affiliation between human and more-than-human develop. The haptic dimensions of the film accent that we are never outside water. Close-ups of rain droplets running across bus windows and boiling water enveloping an egg submerge viewers' eyes. The sounds of leaking pipes and water splashing as a woman masturbates in a bathtub drown ears. In every way, shape, and form, water is pervasive. From waves to ripples, we are immersed by the various shapes of water. Trickles and drops, bubbles and spray, amplify the intimacy between flesh, water, and place.[44] Small-scale bodies of water accentuate inevitable contact with water, and the inherent porosity of our own trans-corporeal nature. Everyone in *The Shape of Water* is saturated and inescapably enmeshed with non-human bodies of water, especially as water leaks into domestic space and dampens military complexes. Dry moments are difficult to find, even in fortified facilities constructed to master life beneath the surface.

Elisa Esposito and Zelda Fuller work in one such facility. They work the night-shift, cleaning floors at a high-security government laboratory near Baltimore, Maryland. It is the height of the Cold War, and researchers are racing to find technology that will help send humans into space. The two women are assigned to clean a room where classified experiments are taking place, experiments with a creature taken from Amazon rivers. For military personnel, the arrangement seems ideal. Elisa is mute, and Zelda is a black woman working in a white space. Both are surrounded by men with institutional authority and voices of patriotic fervor. Silenced women, in the view of Richard Strickland, head of security, will help keep secrets, and hierarchies, intact. Before long, Elisa and Zelda encounter the "asset," an otherworldly sea-creature, described by Strickland as a "filthy thing," dragged "out of the river muck in South America." The monster is no mermaid. Scales and fins, webbed hands and feet, gills and claws are pronounced.

As are the eyes (bluish with hints of green and gray) and the way it blinks. Yet, it stands on two legs, walks, and has hints of human form. As we learn, it cannot be out of water for long. Nor can it morph into a more human form, which Strickland uses to justify why the monster is deserving of disgust and dissection. Addressing Elisa and Zelda, he states,

> *Let me say this up front: You clean that lab, you get out. The thing we keep in there is an affront. Do you know what an affront is, Zelda?*
>
> Zelda responds: *Something offensive?*
>
> Strickland continues: *That's right. And I should know, I dragged that . . . filthy thing . . . out of the river muck in South America all the way here. And along the way we didn't get to like each other much. Now. You may think, "That thing looks human." Stands on two legs, right? But – we're created in the Lord's image. You don't think that's what the Lord looks like, do you?*[45]

With divine certainty, Strickland embodies a Euro-centric conquering gaze, hinged to phallocentric militarism and technocratic ordering. Strickland hinges divine truth to an enlightened scientific gaze, a combination that justifies sentiments expressed by higher-ranking military officers who order Strickland to "cut it open" and glean what knowledge we can.[46] There are dissenters to this kind of gaze. A Russian spy and scientist appeals to authorities by citing that "the asset" displays capacities for emotional intelligence and language.[47] The appeals fall on deaf ears and armored masculinity. The distanced, ordered, and rational gaze of what James C. Scott calls the authority of high-modernism holds too much sway.[48] Given associations with jet engines and a desire for "infinite vision" and an unmarked, disembodied perspective, Strickland disdains fleshy encounters, especially those involving scales and gills. Strickland's response is brutal, bludgeoning enforcement of human/monster binary.

Elisa mops up water saturated with the blood from the creature's taming. Blood—human and non-human—and water spill together throughout the film, accenting how both flow between bodies. Strickland's blood mixes with the creature's. Wet mops cannot soak it all up. We see fluids form impure mixtures on sterile floors, accenting how insides and outsides move together. We see that blood ties and watery bonds are not much different. Keen attention to the flows of blood and water informs Elisa's more epistemologically humble approach to the river being. Embracing Haraway's "situated knowledge," Elisa becomes answerable for what she sees. She cannot sterilize the scene of violence. Nor does she desire an elevated, disembodied subjectivity. Instead, Elisa starts to share intimate moments with the monster, something her muteness makes possible. Elisa's attentiveness to unspoken meaning and gestural acts of communication make affiliation a matter of affect and the ability to be affected. Not reliant on *logos*, Elisa and the creature rely on touch and sound to form a trans-species kinship. As Adriana Cavarero argues, sharing sound and breath is at the heart of meaning making, particularly as it relates to being with "others."[49] Elisa and the creature also share touches of hands on glass and hard-boiled eggs. Moments of sharing are tentative at first, but they quickly turn into a deviant posthuman erotic affair. Soon flesh and scales are entangled in an "erotic-ethical affiliation" between human and more-than-human, a creative symbiosis that brings human bodies and river ecologies into affective proximity.[50]

The relationship motivates a rescue. Elisa and the creature escape the laboratory with the help of a lab researcher, Zelda, and another friend named Giles. It is worth noting that Strickland's conquering gaze is what makes the escape possible. Strickland overlooks Elisa and Zelda as inhuman, dismissing their difference. Zelda is a black woman, and Elisa lacks the capacity for speech. Their status as those who mop up Strickland's bloody messes adds to their perceived inferiority. In attempting to find those responsible for letting the monster loose, Strickland interrogates everyone at the lab, even those he deems incapable of carrying out such an act: "What am I doing, interviewing the fucking help? The shit cleaners. The piss wipers." As it turns out, those closest to shit and piss are the most ethical agents. Being exposed to waste, filth, and bodily substances is what precipitates Elisa's connection to a "filthy" river-creature. A shared "inhumanity" between Elisa, Zelda, and the monster constitutes an inter-species alliance that defies the humanity of Strickland's godly image. This defiance is most explicit in Elisa's responses to Strickland's interrogation. In response to Strickland's demands, Elisa signs "F.U.C.K. Y.O.U.," a message the interrogator cannot understand. A message that he sees as not worth "hearing." Elisa's reliance on hands and touch to communicate might be perceived as a lack of speech, and a lack of humanity, but it is what shapes her posthuman erotic-ethic and the queer attachments that she and the creature have developed. According to Elisa, "When he looks at me, the way he looks at me . . . He does not know what I lack . . . Or how I am incomplete."[51]

Elisa's incompleteness, including her reliance on touch and bodily gesture, animates what Alaimo calls "feminist exposure"—an ethical orientation informed by inevitable vulnerability and porosity.[52] In response to Strickland's constant, masculine attempts to master water creatures and make himself invulnerable against the flux and flow of water, Elisa embraces the deviant pleasures of exposure to watery bodies. The most provocative scenes are where flesh meets scales in naked embrace. Elisa and the creature are seen as exposed to each other, pressed together in non-procreative biophilia—intimate and deeply felt affinities between human and non-human ecologies.[53] For Elisa, it is a case of aquaphilia that leads her to seal off her bathroom and flood the space with water. The flooded room allows an erotic encounter. Here, built domestic spaces become watery, stressing the fluid ways we are connected to aquatic environments. The small domestic space is not waterproof. Water drips into apartments below, saturating wooden walls. The entire space becomes a body of water. Elisa shows us that we need not go out to sea to feel proximity to marine life. Rather, it is a matter of recognizing how porous intimate spaces are. When the door to Elisa's bathroom opens, water spills out to reveal sea creature and Elisa entangled with each other, a "thoroughly perverse, and wildly anarchic" vision of biophilic composition.[54] Elisa looks over the creature's shoulder with an expression of defiant pleasure, exhibiting a queer ecological love that challenges the "biophobic moralizing" that keeps bodies of water and human bodies distant and distinct.[55] An ocean of bodily fluids mix in the apartment as a promiscuous act of aquaphilia triumphs over the phallocentric military-bio-research complex. Touching in water provides a pleasurable ethic in the face of batons wielded by men hell bent on taming and maintaining the mastery model of anthropomorphism.

Elisa's freakish experiment with a river creature offers a "queer paradigm of desire" that embraces irreverent, "creative variations" of ethical affinities that stress affective proximity.[56] Importantly, the erotic ethic Elisa practices does not require humanizing, or making human. Scales, fins, and webbed appendages are enough for Elisa, who knows all too well the violence that ideal humanism motivates. Elisa finds the fishy elements beautiful, even stimulating, perceptions that risk her own humanity. Affirming affectability, Elisa embraces the radical exposure to aquatic life that Alaimo sees as necessary for developing more ethical relations with seas and sea-creatures. Elisa's radically erotic embrace animates an ethical orientation devoted to the queer exuberance of ecology, the inherent sexual variation and plasticity beneath the surfaces of what we call "nature."[57] The posthuman ethic promoted in *The Shape of Water* exposes us to the non-linear shape of becoming and the potential for creative, sympoetic ways of becoming with water.[58] There is an affirmation of queerly sexualized nature and promiscuity in natural relations, all of which reminds us about our own embodied naturalness— or rather, our own swampiness, and the possibilities of queer aquatic affiliations that complicate, elaborate, divert attachments to aquatic life.[59] Elisa's embodied transgression of the human/river-creature binary evinces malleability and diversity in the ways we might feel out the shape of water and the shape our water is in. With Elisa, we might even learn to love it.

Speculating ethics at the edge of the shore

> Unable to perceive the shape of you, I find you all around me. Your presence fills my eyes with your love, it humbles my heart. For you are everywhere.
>
> – Giles, *The Shape of Water*

The latest images emerging from Florida's red waves of algae toxins are what some would call horrific. Unprecedented numbers of poisoned sea life float to the surfaces and wash ashore. It is a vortex of death. Microscopic organisms produce brevetoxins that impact nervous systems, while waves and wind carry poison air into human lungs. Though toxic algae bloom is not uncommon along Florida gulf coastlines, this one is expected to last well beyond typical timelines. Intensified by agricultural runoff and high temperatures, the Red Tide has been declared a state of emergency. It is difficult to imagine pelagic love flourishing in such conditions. In fact, it is safe to say that these toxic tides kill any chance of radical aquaphilia. Standing at the edge of Florida's water, we are left looking for sea creatures yet to come, the ones who might inspire more loving affinities and submersive and subversive aquatic affiliations. If it's not too late. If we haven't left our monstrous kin to die beneath toxic tides and by-products of earthy agri-logics. Alternatively, it may be that the algae bloom produces and/or awakens creatures from the deep that have no choice but to surface and demand response. How will we, how should we, welcome them to the shore? Will we enclose them in concrete military bio-technical research facilities? If so, will we help them escape? Or is it possible to encounter them in more dynamic ways—as guides into new affinities with water? Is it possible to understand them as leading us to a more aquatic sense of self, culture, and ecology? Are we willing to take a plunge?

At the end of *The Shape of Water*, we are left at the edge of dry land, at a point of judgment. We are left to decide whether or not to dive in with monsters and expose our human bodies to bodies of water. We end where this essay began. Elisa and the river creature stand at the edge of a harbor, in the pouring rain, saying goodbye. Elisa is under no illusion that a life under the sea is possible. Before they can part, however, Strickland arrives to recapture his monster and ultimately put an end to the transgressions its presence has prompted. Drawing a revolver, he shoots the creature three times in the chest before shooting Elisa once in the stomach. The lovers fall, but the monster rises. Confronting a shocked Strickland, the creature swipes at the shooter's throat, leaving the masterful man bleeding and dying on the concrete. Strickland's blood flows with rain water into the sea, once more reminding us how pelagic our blood is. In the end, even Strickland becomes a watery body. Returning to Elisa, the creature picks her up, holds her close, and jumps into the water. Once below the surface, things morph. Three scars on Elisa's neck transform into gills, suggesting that Elisa is returning to her own watery origins. While mute on land, water provides her with voice. She is where she belongs. This is not a little mermaid tale, in which land is *the* place where the mermaid wants to live. Instead of a creature losing fins to live among men, a woman gains gills for an aquatic life, an instance that emphasizes how human bodies are "sea, sands, corals, seaweeds, beaches, tides, swimmers," waves, and sponges.[60] That we are permeable bodies caught up in the flow of ecologies and economies.[61] That distinctions between monsters and humans are untenable, and that human bodies, minds, and souls are fundamentally shaped and shifted by material immersions. Elisa's metamorphosis suggests that it is possible to change who we are in relation to oceans. It is possible to allow oceans to change who we are.

However, what are we to become as watery bodies? That depends on the shape of our waters. Oceans are becoming more acidic and will be home to more plastic than fish by the year 2050.[62] The shape of our waters are toxic, suggesting that we are already turning into "monsters," creatures shaped by organic and inorganic material, by salt and decomposing straws. The ocean Elisa finds is not vibrant. It is not a panacea to life on land. The final shot of *The Shape of Water* leaves us with an image of Elisa and the creature surrounded by blurred blue-black waters. Rather than an ocean teeming with life, we are submerged into a dark aquatic ecology, left not knowing what kind of ocean we are in.[63] While the open water might provide Elisa with rebirth, escape, and reclamation of watery origins, we are uncertain what she is swimming into. The un-determined status of these depths "throws water on the happy dreams of environmentalism" and visions of pure watery politics.[64] In the end, we don't fully know what happens to Elisa and the monster. The camera fades out as water takes them into the abyss. It may be that the un-determined end of *The Shape of Water* asks us to reconsider what kinds of waters are required for sustaining a loving embrace with sea life.

Falling in love with monsters and cultivating attraction to sea life can mean radical shifts in how we imagine our permeable relations to water. That is, as long as we can welcome strangers from the deep as figures that are both threatening *and* generative, in that they threaten dominant categories, "hermetic paths," and boundaries of eco(nto)logical difference, while generating tension at the boundaries of what is possible, prompting desire at the "thresholds of becoming."[65] Monsters "ask us to consider the wonders and terrors of entanglement" with what emerges in the ruins of capitalism and the uneven effects of human transformations of multispecies life.[66] Mermaids and river creatures lead us towards thresholds, towards the wonder and terror of living in the pelagiocene. To be sure, most of us don't have a creaturely guide to life below surfaces. The most we can hope for are moments of submersion, furtively grasping for breath. Even with borrowed organs and submersibles, full immersion and complete knowledge are impossible. However, there is, just below the surface, a kind of breath, between inhalation and exhalation. Neimanis calls it a pause, a point at which human lungs are locked between breathing in and out. The formal name for this state is *aspiration*, suspended between water and air. Even if this is the closest we get to "becoming fish," such in-between moments provide a sense of what it might mean to *become with* our strange aquatic kin.[67]

Importantly, even as we might be "land-locked," we might find moments of aspiration in speculative fiction, contemporary sea stories inspired by Noah and Odysseus, Jacques Cousteau and Calypso. Sirens continue to surface in popular culture, suggesting that we are in a tentacular moment.[68] It is no coincidence that sea monsters populate contemporary popular culture. As I compose this sentence, stories of a juvenile whale washing ashore in the Philippines circulates around digital media.[69] Its stomach was filled with nothing but plastic, a monstrous creation in itself, one we should fear, even as it might press us to engage with the mundane violence in the detritus of capitalism. These very real monstrous blends of human and more-than-human elements demand a posthuman ethics that embraces deeper affinities with those that lurk in the hadal zone, those that remind us how "our own bodies harbor not only watery traces of evolutionary pasts but also the latent watery potential of . . . futures not chosen,"[70] futures in which we embrace and reimagine oceanic selves and our relations with strange, monstrous oceanic kin.

Notes

1 "Life worlds" is a phrase used by biologist Jakob von Uexküll, who argues that any biological study must take into account how micro ecologies shape species perception and practice. For Uexküll, there is no essential species, only life worlds of interdependence. See, Jakob von Uexküll, *A Foray into the Worlds of Animals and Humans*, trans. Joseph D. O'Neil (Minneapolis, MN: University of Minnesota Press, 2010).

2 Carey Wolfe, *What Is Posthumanism* (Minneapolis, MN: University of Minnesota Press, 2009).

3 Amanda Nichols, "Being Human: An Ecocentric Approach to Climate Ethics," in *Climate Change Ethics and the Non-Human World*, eds. Brian Henning & Zack Walsh (2020).

4 Val Plumwood, *Feminism and the Mastery of Nature* (New York: Routledge, 1993).

5 Rosi Braidotti, *Metamorphoses: Towards a Materialist Theory of Becoming* (New York: Polity Press, 2002).

6 Nancy Tuana, "Viscous Porosity: Witnessing Katrina," in *Material Feminisms*, eds. Stacy Alaimo and Susan Hekman (Bloomington, IN: Indiana University Press, 2008).

7 Sam Mickey, "Atmospheres of Object-Oriented Ontology," in *Climate Change Ethics and the Non-Human World*, eds. Brian Henning and Zack Walsh (2020).

8 Ibid.

9 Ibid.

10 Donna Haraway, *Staying with the Trouble: Making Kin in the Chthulucene* (Durham, NC: Duke University Press, 2016).

11 Astrida Neimanis, "Hydrofeminism: Or, on Becoming a Body of Water," in *Undutiful Daughters: Mobilizing Future Concepts, Bodies and Subjectivities in Feminist Thought and Practice*, eds. Henriette Gunkel, Chrysanthi Nigianni, and Fanny Söderbäck (New York: Palgrave Macmillan, 2012).

12 Stacy Alaimo, *Exposed: Environmental Politics and Pleasures in Posthuman Times* (Minneapolis, MN: University of Minnesota Press, 2016), 124.

13 Ibid., 5.

14 Ibid.

15 Weeki Wachee Springs State Park is an hour's drive North from Tampa, where I used to live. See, https://weekiwachee.com/.

16 Sole-Smith, Virginia, "The Last Mermaid Show," *The New York Times* (July 5, 2013), www.nytimes.com/2013/07/07/magazine/the-last-mermaid-show.html.

17 For example, see Melissa Bryer, "Meet the Mermaid Trying to Save Florida's Ailing Springs," *Treehugger* (February 1, 2018), www.treehugger.com/clean-water/meet-mermaid-trying-save-floridas-ailing-springs.html; Funes, Yessenia, "The Real-Life Mermaid Fighting to Save Florida's Disappearing Springs," *Earther* (January 31, 2018), https://earther.gizmodo.com/the-real-life-mermaid-fighting-to-save-floridas-disappe-1822568735

18 Jeffrey J. Cohen, "Monster Culture (Seven Theses)," in *Monster Theory: Reading Culture*, ed. Jeffrey Cohen (Minneapolis, MN: University of Minnesota Press, 1996).

19 Luce Iragaray, *Marine Lover: Of Friedrich Nietzsche*, trans. Gillina C. Gill (New York: Columbia University Press, 1991).

20 Alaimo, *Exposed*.

21 Ibid., 22.

22 Donna Haraway, "Anthropocene, Capitalocene, Plantionocene, Chthulucene: Making Kin," *Environmental Humanities* 6 (2015): 159–65. In complicating what has been called the Anthropocene, Haraway notes that assigning culpability to "humans" flattens how "we" are implicated in global climate change. Not all human bodies carry the burden equally, not all reap the economic "rewards" equally. To say that current epochs of ecological devastation are "human" elides important differences in power and the different scales of human participation in clear-cutting, production of toxins, and mass-produced death. I was recently reminded that the East India Trading Company worked to develop the most efficient and direct routes for trade, which certainly required cordoning off directions and killing what got in the way. Operating to guarantee efficient movement contrasted with the ways in which whalers (and other seafaring projects) wandered. Following whales, whalers surely encountered a more strangely diverse seascape.

23 Eric White, "'Once They Were Men, Now They're Land Crabs': Monstrous Becomings in Evolutionist Cinema," in *Posthuman Bodies*, eds. Judith Halberstam and Ira Livingston (Bloomington, IN: Indiana University Press, 1995), 245.

24 Mick Smith, *An Ethics of Place: Radical Ecology, Postmodernity, and Social Theory* (New York: SUNY Press, 2001). In addition to popular culture renderings that animate vibrant images of oceanic life, I am drawn to Claire Nouvian's photographic collection *The Deep: Extraordinary Creatures from the Abyss* and William Beebe's *A Half Mile Down*.

25 Philip E. Steinberg, *The Social Construction of the Ocean* (Cambridge, England: Cambridge University Press, 2001). Of course, going *20,000 Leagues under the Sea* and into *The Core*, for example, challenges such horizontal orientations. There are any number of films that venture beneath surfaces, all of which might throw into relief the fact that the more-than-human flourishes in the strangest places.

26 As Stacy Alaimo notes, sensing what might flow beneath the surface requires economic access and technical expertise, making knowledge about the majority of oceanic ecologies and creatures difficult to obtain. The need for expensive equipment and extensive temporal commitments make it difficult for us to "see" beyond the initial layers of the sea. Thus, speculative storying is necessary.

27 Jeffrey J. Cohen Lowell Duckert, "Eleven Principles of the Elements," in *Elemental Ecocriticism Thinking with Earth, Air, Water, and Fire*, eds. Jeffrey J. Cohen and Lowell Duckert (Minneapolis: University of Minnesota Press, 2015), 12. Entanglements between human and more-than human are described by Stacy Alaimo as transcorporeal, in which porous bodies secrete and absorb, spill into each other. We might also say that the messy "nature" of worlds is composed in terms of what Karen Barad calls "intra-action." In this version, there are no pre-supposed identities, formations, or assemblages that interact. Rather, bodies and relations are formed in action. See, Karen Barad, "Quantum Entanglements and Hauntological Relations of Inheritance: Dis/continuities, Space Time Enfoldings, and Justice-to-Come," *Derrida Today 3* (2010): 240–68.

28 Haraway, *Staying with the Trouble*.

29 Morton, Timothy, *The Ecological Thought* (Cambridge, MA: Harvard University Press, 2010).

30 Michael Hauskeller, Curtis D. Cabonell, and Thomas D. Philbeck, *The Palgrave Handbook of Posthumanism in Film and Television* (New York: Palgrave Macmillan, 2015).

31 Linda Nash, *Inescapable Ecologies: A History of Environment, Disease, and Knowledge* (Berkeley, CA: University of California Press, 2007).

32 Tuana details how flood waters and flows of toxins following Hurricane Katrina seep through porous skin, turning human bodies into much more than human bodies of poisoned water.

33 This sentiment is only amplified by authors of Ebook erotica that feature tentancled relations. See the provocative essay by Dagmar Van Engen, "How to Fuck a Kraken: Cephalopod Sexualities and Non-binary Genders in Ebook Erotica," *HUMaNIMALIA 9* (2017): 121–151.

34 Steve Mentz, *At the Bottom of Shakespeare's Ocean* (New York: Bloomsbury Press, 2009), 98.

35 Haraway, *Staying with the Trouble*, 52.

36 Ibid., 55.

37 Ibid., 56.

38 Ibid.

39 Cohen and Duckert, "Eleven Principles," 20.

40 Judith Halberstam, *The Queer Art of Failure* (Durham, NC: Duke University Press, 2011). Doing so aligns with Judith Halberstam, who argues that engaging with

dismissed cultural artifacts as "low theory" can animate creative critical/intellectual/ activist work. Low theory provides opportunities for messy mis-readings, productive when trying to alter social imaginaries and material relations. Halberstam's queer theory favors "sinking" lower, an attitude most relevant for a project focused on watery realms. In this chapter, then, some theoretical highlights will sink into footnotes.

41 Alaimo, *Exposed*, 158.
42 Astrida Neimanis, *Bodies of Water: Posthuman Feminist Phenomenology* (New York: Bloomsbury Press, 2017), 137.
43 Ibid., 146.
44 Alaimo, *Exposed*, 77.
45 As the conversation continues, Strickland makes it clear that "some of us" are more like God than others, stressing that black women are closer to creature than the divine. God, then, is seen in Strickland.
46 The trouble with the scientific gaze is described by Donna Haraway, who notes that scientific visions construct relations, even as they might claim to simply reveal. See, Donna Haraway, *Primate Visions: Gender, Race, and Nature in the World of Modern Science* (New York: Routledge, 1990).
47 The argument about language as a distinctly human province, as something that makes humans uniquely unique, has been challenged by philosophers, animal biologists, and communication/rhetoric scholars, who are rethinking the limits of rhetoric—what it is, who does it, and how posthuman rhetoric is expansive and ecological. See, for example, McGreavy, Bridie, Wells, Justine, George McHendry, and Samantha Senda-Cook, *Tracing Rhetoric and Material Life: Ecological Approaches* (New York: Palgrave MacMillan, 208); Emily Plec, *Perspectives on Human-Animal Communication: Internatural Communication* (New York: Routledge, 2012).
48 James C. Scott, *Seeing Like a State: How Certain Schemes to Improve the Human Condition Have Failed* (New Haven, CT: Yale University Press, 1999).
49 Adriana Cavarero, *For More Than One Voice: Towards a Philosophy of Vocal Expression*, trans. Paul A. Kottman (Stanford, CA: Stanford University Press, 2005).
50 Dianne Chisholm, "Biophilia, Creative Involution, and the Ecological Future of Queer Desire," in *Queer Ecologies: Sex, Nature, Politics, Desire*, eds. Catriona Mortimer-Sandilands and Bruce Erickson (Bloomington, IN: Indiana University Press, 2010).
51 Though time and space limit a rich discussion of the intersections of posthuman erotic-ethics and disability studies, future research will provide for in-depth attention to how other-abled approaches to ecology are critical for considerations of climate ethics.
52 Alaimo, *Exposed*.
53 E.O. Wilson, *Biophilia: The Human Bond with Other Species* (Cambridge, MA: Harvard University Press, 1984).
54 Chisholm, "Biophilia," 377.
55 Ibid.
56 Ibid., 376.
57 Bruce Bagemihl, *Biological Exuberance: Animal Homosexuality and Natural Diversity* (New York: MacMillan Publishers, 2000); Joanna Roughgarden, *Evolution's Rainbow: Diversity, Gender, and Sexuality in Nature and People* (Berkeley, CA: University of California Press, 2009).
58 Myra Hird discusses how non-linear evolution brings queer theory and evolutionary theory together to fashion a "naturally queer" ecology. See, Myra J. Hird, "Naturally Queer," *Feminist Theory* 5 (2004): 85–89.
59 David Bell, "Queernaturecultures," in *Queer Ecologies: Sex, Nature, Politics, Desire*, eds. Catriona Mortimer-Sandilands and Bruce Erickson (Bloomington, IN: Indiana University Press, 2010), 331–58.
60 Helene Cixous, & Catherine Clement, *The Newly Born Woman*, trans. Betsy Wing (Minneapolis, MN: University of Minnesota Press, 1986).

61 Braidotti, *Metamorphoses*.

62 Graeme Wearden, "More Plastic than Fish in the Sea by 2050, Says Ellen MacArthur," *Guardian* (January 19, 2016), www.theguardian.com/business/2016/jan/19/more-plastic-than-fish-in-the-sea-by-2050-warns-ellen-macarthur.

63 Timothy Morton, *Dark Ecology: For a Logic of Future Coexistence* (New York: Columbia University Press, 2016).

64 Mentz, *At the Bottom of Shakespeare's Ocean*, xii.

65 Cohen, "Monster Culture (Seven Theses)," 7, 20.

66 Anna Tsing, Heather Swanson, Elaine Gan, and Nils Bubandt, *Arts of Living on a Damaged Planet* (Minneapolis, MN: Minnesota University Press, 2017), M2.

67 Vera Coleman, "Becoming a Fish: Trans-Species Becomings in Narrative Fiction of the Southern Cone," *ISLE: Interdisciplinary Studies in Literature and Environment 23* (2016): 694–710. Kin that are getting even stranger. See, Signe Dean, "Octopus and Squid Evolution Is Officially Weirder than We Could Have Ever Imagined," *Science Alert* (March 17, 2018), www.sciencealert.com/octopus-and-squid-evolution-is-officially-weirder-than-we-could-have-ever-imagined.

68 I am referencing the rise of shows like *Siren and Tidelands*, both of which feature the arrival of sirens and complex relations between land and sea. Both were released in 2018.

69 Hannah Ellis-Peterson, "Dead Whale Washed Up in Philippines Had 40kg of Plastic Bags in Its Stomach," *Guardian* (March 18, 2019), www.theguardian.com/environment/2019/mar/18/dead-whale-washed-up-in-philippines-had-40kg-of-plastic-bags-in-its-stomach?fbclid=IwAR0e43hG0Ty_c-vKJEPAQq4CS81vltpPByZDNBImJJd_vLeM--bLMuHiHyI.

70 Ibid., 135.

13 Gut check

Imagining a posthuman "Climate"

Connie Johnston

At the scale of both the global climate and the microbial, in this chapter I would like to explore and hopefully disrupt ideas about "anthropo." "Anthropo," of course, is the Greek root for our species or things related or similar to us. The geologic age we live in has been labelled the "Anthropocene" because of the scale of (frequently devastating) change that our species has created upon the Earth. It is not false that we humans as a collective species have had a profound impact on other species and the Earth's functions. A recent study, in fact, states that the melting of polar ice caused by climate change is actually having an effect on the Earth's rotation (Mitrovica et al. 2015). Relative to our planet's existence, these impacts have been exceedingly swift because of the global scale of human activity and pace of technological development. Through a variety of media, we can observe manifestations of climate change on an ongoing basis: melting glaciers and polar ice, repeated and worsening coastal flooding, and horrific wildfires. The effect on human and non-human life embedded in these environments is easy to see (witness images of polar bears literally and figuratively on thin ice). However, visible life forms are themselves environments for the microscopic. Florae inhabit the intestinal tracts of all animals, and studies are showing that the effects of climate change may even be reaching into these most intimate of spaces.

In this chapter, I will first review recent findings related to the environment, climate, and intestinal flora. I will then discuss the idea of the unitary human subject. Next, I will pose several challenges to the concepts of "anthropo" and the human subject. Then I will review key components of posthumanist theory in order to, finally, explore how we might envision a posthuman "climate." Climate change will change who "we" are, but likely in unexpected ways. Using posthumanist theory, I will claim (following Braidotti 2013 in particular) that in order to truly shift an anthropocentric ethical perspective, a challenge to the concept of the human self is necessary.

The environment and gut flora

It is becoming more widely and popularly known that scores of microorganisms inhabit our human intestinal tracts (as is the case for all animals with intestinal tracts) and that these organisms—our "gut microbiome"—play an important role in the proper functioning of our bodies (Bolt 2015). Recent studies are indicating that these florae respond not just to foods eaten but also to the body's exterior environment. For example, researchers at the University of British Columbia state that comparative studies show that aspects of typical Western lifestyles lead to a decrease in gut microbiota diversity. These aspects are such things as antibiotic use, sanitized living environments, and reduced breast feeding of infants. The researchers note that the environment is as key a factor as the mother in transmitting beneficial microbes to infants and that children who grow up in rural environments show more gut microbial diversity than those who do not. Based on this research, but lacking sufficient comparative data at this point, the authors also suspect a relationship between soils in different regions and the gut flora of the regions' residents (Tasnim et al. 2017). A study by researchers at Stanford University showed seasonal variation of the gut microbiome in Hazda hunter-gatherers in Tanzania; the researchers also compared the microbial taxa with those found in industrialized populations. Although the Hazda diet changes with their distinct wet and dry seasons, the authors noted that the gut microbial diversity in the Hazda population was, overall, higher than the diversity in the gut of those in more industrialized societies, indicating that both dietary fluctuations and the environment shape the gut microbiome (Peddada 2017; Smits et al. 2017). And a recent, small study by scientists at the Center for Microbiome Informatics and Therapeutics showed that diet may have less of an influence on a person's gut flora than would be expected. Despite the research participants' being kept on an identical diet for the six-day study period, daily fluctuations in gut flora were still experienced (Gurry et al. 2018).

Although all of the authors of the preceding studies stated that more research is needed to study gut microbiomes in general, and the effect of the environment on them in particular, each publication concluded that an organism's external environment plays a potentially greater role on the makeup of gut flora than has been assumed. In particular, the first study reviewed (by Tasnim et al.) included a section that speculated on the relationship between local soils and inhabitants' gut flora, with concerns expressed over soil degradation and the current and future effects of climate change. Another recent study, however, explicitly examined the effects of a changing (specifically, warming) climate on gut flora. Bestian et al. (2017) researched temperature changes' effects on the gut flora of the common lizard (*Zootoca vivipara*). Overall, the study found that warmer temperatures had a negative impact on the lizards' gut microbiome in terms of species richness. Based on the wide range of functions in which these microbial organisms play a (often poorly understood) role, the authors state that their results indicate that negative effects could span the health and survival of hosts and, thereby, have an impact on larger ecosystems within which the hosts are situated. Although the scientists note that, to date, the effects (and routes of those effects) of climate change on gut bacterial communities are not well understood, they do state,

> One of the most intricate symbiotic relationships is probably that between animal hosts and the bacterial community inhabiting their guts. The gut

microbiota is shaped by host traits ... and environment ... and, in turn, plays multiple essential functions for hosts, including digestion, immunity and life history. Consequently, changes in the gut microbiota could lead to potential dysbioses with strong consequences for hosts and ecosystems. These changes may be caused by rising temperatures that lead to an increase in disease prevalence.

(1)

What the studies reviewed here tell us is that, in broad terms, there is another potentially cascading effect of climate change—one that travels on multiple pathways from the "outside world" into that intimate space of an animal's gut. This impact of climate change would clearly have effects that cut across innumerable species, both human and non. Before I delve further into this space of the gut, however, I would like to pause and talk a bit about the idea of the human subject.

The individual human (subject)

As stated earlier, a main goal of this chapter is to pose an ontological challenge to "anthropo"—the human. To do this, we need to not take "anthropo" as a material-conceptual given, but to de-materialize it and review the origins of "the human" as an idea. In an article on the relationship between climate change and human conceptualization of time, historian Dipesh Chakrabarty (2009) states that the development of the idea of "the human" was essential in the creation of an object of analysis in various Western social contexts. These contexts—historical, biological, medical, and so on—and therefore the precise composition of "the human," may differ, but this entity is still a reduction or abstraction, regardless. In other words, the figuration of the human varies depending on how the concept will be used, but no representation can capture the totality of characteristics that may reasonably be identified as human. Therefore, no individual, material "human" organism will ever fully correspond to any one conceptualization.

Although not explicitly drawing on Agamben (2004), Chakrabarty's assertion that "the human" is a concept of utility, rather than an exact representation of a material entity, in many ways links with Agamben's theorization of "the anthropological machine." For Agamben, the production of "the human" is rooted in the work of Aristotle and is foundational in Western scientific, intellectual, and socio-political developments and institutions. Aligning with the Western conceptual tendency toward binaries, this concept of "the human" posits a split between the human and animal, isolating the purely biological animal within the socio-cultural being of the human. This division works to create the rational human subject upon which Western institutions (continue to) rest. The concept of the rational human subject, capable of culture and reason, is, importantly, oppositional to the biological animal or the "not human" (although numerous human groups throughout the history of Western society have also been placed in the "not human" category). The reification of "the human" reflects not materiality but an ideal and also a tool for creating and justifying power.

Agamben's interrogation of the foundations of Western institutions is an important reference point for a number of posthumanist scholars (e.g., Braun 2004a, 2004b; Lorimer 2009; Braidotti 2013), and it is from this area of scholarship[1] that much fruitful analysis of the construction of the human is done. In "Mapping Posthumanism" (2004), Neil Badmington highlights the grip that the human subject still has on Western thought today. He specifically provides a critique of political scientist Francis Fukuyama's 2002 *Our Posthuman Future: Consequences of the Biotechnology Revolution*. Here, Badmington asserts that Fukuyama is fearful of the potential "loss" of the human and tries (unsuccessfully) to define "human nature." Ultimately, according to Badmington, Fukuyama's arguments reflect "[h]umanism [that] is an obstacle, an ideology, a myth" (1349). In a similar vein, Castree and Nash (2006) point to ongoing humanistic thought and its "investment" in the constructed idea of "the human." Emphasizing the non-universality of the "universal human," Lorimer (2009) states that the deconstructive impulse of posthumanism calls out this ideological construct that has been figured as Western, white, male, and heterosexual.

What has arisen out of Western society's humanistic past, and continues into the present, is the idea/ideal of the unitary, universal human subject, or what we frequently think of as "the individual." Posthumanist scholarship has interrogated not only the creation of this entity but also what its characteristic traits are and how the concept is maintained. As Braun (2004a) indicates, posthumanist scholarship plays two main roles, one related to deconstructing the unitary human subject and the other to theorizing and "suggesting" alternative ontologies. Posthumanism indicates a perspective, rather than an historical moment (Castree and Nash 2006), one that seeks to disrupt humanism's "ontological hygiene," or the careful maintenance of conceptual boundaries "separating human from non-human, nature from culture, organism from machine" (Graham 2002, 35). Humanism, its ideals, ideologies, and constructs are sufficiently persistent, however, such that they will not be erased by one prefix.

A concern evident in much of the critical scholarship feeding into posthumanism (e.g., Butler 1993; Foucault 1980; Derrida 2008; Said 2004; Haraway 1991) and posthumanism itself relates to ethics. Specifically, how has this ideal of the universal human subject that has persisted in Western culture had an impact on earthly life? Certainly, feminist and postcolonial scholarship has taken this figure to task, and volumes have been written about the violence done to humans who are deemed not to "measure up" to this ideal. Violence here indicates not just direct physical violence, although that certainly occurs, but also the more indirect violence of institutions that are based on the ideal of this particularly configured autonomous individual (Whatmore 2002; Lorimer 2009). With respect to the non-human world, the feminist philosopher Val Plumwood (1993) has highlighted the "imperialism of the self," which mandates an ethical framework based on this figure of the individual, thereby limiting different conceptualizations of ethical relationships. Because the concept of the human, as a rights-bearing, individual self, has been problematic, by extension the (Western) ethical concept of rights has also been problematic. One of the projects of posthumanism is to break away from these limitations that are based on classical humanist ethical frameworks deriving from Kantianism or utilitarianism (Braidotti 2013). Following Agamben's ethical project in *The Open* (2004), rather than seeking the "better" version of an already flawed ethical framework that can produce violent outcomes, perhaps we should be determining how the anthropological machine "work[s] so that we might, eventually, be able to stop [it]" (38). As a first step in this project, in the next section, I want to put forward two broad ontological challenges to "anthropo," the subject.

Challenges to "anthropo" and the subject

My first ontological challenge to "anthropo" is temporal (although it does have a spatial element as well). I want to question the acceptance of the idea of the "Anthropocene." With respect to climate change, the key component of the call to arms in this fight is that the changes are as a result of human activity, or anthropogenic. The Earth's climate overall has "changed" before, and therefore large-scale climate shifts are not a new occurrence. For example, although its existence and extent are somewhat subject to debate, there is evidence for the Medieval Warm Period, and the effects on agriculture of the "Little Ice Age" that followed this period are well-studied (Mann 2002). However, as the now-famous "hockey stick" graph[2] in Al Gore's *An Inconvenient Truth* (Guggenheim 2006) illustrated, the general warming of the planet has continued to sharply rise since the advent of Global North industrialization. Rightfully so, therefore, the current global climate crisis is attributed to human activity, as the Intergovernmental Panel on Climate Change has concluded, stating in their most recent report that "human influence on the climate system is clear and growing, with impacts observed across all continents and oceans" (IPCC 2014, v).

Although we can consider our naming of the current planetary age as a desire to accept responsibility for the climate cataclysms we have set in motion, there is nonetheless the ring of arrogance to that naming as well, and the "Anthropocene" reinscribes "anthropo" as the center and driver of earthly events. To counter this temporal anthropocentrism, I reference the late evolutionary biologist Stephen J. Gould (1996), who, two decades ago, pointed out the hubris in the name, the "Age of Man" (sic), stating that our planet is and has always been in the "Age of Bacteria." Of primary importance in disrupting our species' temporal hegemony is the incomparable longevity of the tiniest life forms, estimated to have begun their tenure on Earth about 3.5 billion years ago. Multicellular life forms did not show up until approximately 580 million years ago and, depending how far back in humans' lineage we want to go, our ancestors did not appear until, at the earliest, approximately ten million years ago. Our planet had already lived most of its life by the time our more recent ancestors arrived, and microbes were, of course, still hanging around. Another aspect of temporality is another quality of microbial life—indestructibility. As Gould states, "The [bacterial] organisms cannot be nuked into oblivion or very much affected by any of our considerable conceivable malfeasances" (p. H01). Gould also cites microbial taxonomic diversity, sheer numbers, and, related to spatiality, their ongoing ubiquity on planet Earth. One can argue against Gould's points with respect to humans' relative impact, based on our fewer numbers than microbes, and also what we have been able to accomplish in our shorter time on the planet. However, one cannot argue that we are, from a microbial perspective, more of a "flash in the pan."

My second ontological challenge is primarily spatially oriented, or a challenge to the "anthropo-scene," and I want to start with an extraordinarily brief review of several key moments in Western history that have to some extent decentered the human species. The first is the familiar story of Copernicus's theory, from the 16th century, that replaced the existing geocentric solar system with our current heliocentric one. Perhaps Earth was/is still the only planet in our solar system to have life, but after Copernicus, humans and our home base came to be recognized as just another of the Milky Way's planets, all revolving around the same big star.

The second key moment is, of course, Darwin's publication of *On the Origin of Species* in 1859. Darwin, as is well known, did not construct the theory of evolution but, with his theory of natural selection,[3] provided the mechanism for how evolution (and, thereby, speciation) could actually work. With Darwin, a plausible natural mechanism had been put forward for the way in which Earth's species, including humans, came to exist. Although it may have been (and still is for some) desirable, it was no longer *necessary* for us to have been the special creation in the image of a divine entity. It became entirely possible that humans were just one of many forms of life on Earth, arising in the same way, with our own physiology reflecting that of other living beings.

The final key moment, for my purposes here, is the relatively recent change in the Linnaean kingdoms of life that were developed by the Swedish botanist, physician, and zoologist Carl Linnaeus in the 18th century. Although challenges have been posed to the Linnaean classificatory system (e.g., Schiebinger 2008) and the system has evolved, the Animalia and Plantae kingdoms had persisted as the highest levels of classification until the 1990s. Although still disputed in some quarters, US biologist Carl Woese's proposed three broad "domains"— Archaea, Bacteria, and Eukaryote—have replaced the two kingdoms. Humans' residence—Animalia—is subsumed as one of six branches (which include plants and fungi) of multi-celled organisms under Eukaryote (Johnston 2017). Like our planet and our place in the presumed hierarchy of animal life on Earth, the "kingdom" in which we previously resided has been demoted to a branch on the tree of life.

Linking with this final historical challenge is another to the "anthropo-scene"— the microbial residents of animal guts. Although the processes and tempos of eating and digestion vary by species, all animals must sustain themselves through consuming and then processing what is consumed in some way. This, at the most basic level, connects us as humans to other planetary life. The gut microbial role in bodily sustenance is a more particular part of that connection for animal species. To exist as a living being (however categorized), at the most basic level, involves these processes, and, therefore, these tiny organisms that facilitate sustenance establish a fundamental connection across life forms.

As briefly noted at the beginning of this chapter, intestinal flora plays a role in the proper functioning of the body overall, not "just" in proper digestion. Studies are increasingly pointing to the wide range of issues and processes that gut flora impacts, including immune responses, the development of diseases and allergies, and, important for the points I want to make next, neurological functioning (Jandhyala et al. 2015; Thursby and Juge 2017). For a number of years now, there has been speculation that there are connections between human behavior and the gut and its bacteria. In fact, in a 2014 article in *The Journal of Neuroscience*, Mayer et al. call this new neuroscientific focus a "paradigm shift," stating,

> The discovery of the size and complexity of the human microbiome has resulted in an ongoing reevaluation of many concepts of heath and disease, including diseases affecting the CNS [central nervous system]. . . . The initial skepticism about results suggesting a profound role of an intact gut microbiota in shaping brain chemistry and emotional behavior has given way to an unprecedented paradigm shift in the conceptualization of psychiatric and neurologic diseases.
>
> (p. 15490)

Whereas this line of research is exploding in the present day, the neurological role of the gut was explicitly recognized two decades ago by Michael Gershon, MD, in *The Second Brain* (1998). This "enteric nervous system" of course regulates digestion but performs many more complex actions as well, a number of which are only beginning to be understood in the relatively new field of neurogastroenterology (Hadhazy 2010). Autism has received much of the recent attention as a brain/behavioral disorder with likely gastrointestinal connections (Moyer 2014; Li et al. 2017), with studies indicating that gut microbiomes can be significantly different in people with autism (Kohn 2015). In addition to the autism/gut connection, there is also research focused on other brain and behavioral issues, and scientists at a variety of institutions are receiving large grants to investigate the relationship between gut bacteria and such things as brain development, schizophrenia, depression, and anxiety (Elliott 2013; Bolt 2015; Kohn 2015; Autism Speaks 2018).

All of the preceding should contribute to a destabilization of the, at least in Western society, defining feature of "anthropo"—the capacity for reason, seated in the brain. Further, how we conceptualize "the individual" and the characteristics that make a person who she/he/they are, are tied up in the types of things that gut microbes are increasingly being shown to impact. Do we describe ourselves or others as optimistic, melancholic, intelligent, well-adjusted, and so on? These qualities may have less to do with the individual human "self" than we have wished to believe. What I am suggesting here is that the studies referenced herein (and this burgeoning field overall) are pointing to exploding the notion of the unitary, rational human subject. It seems that what makes us "who we are" is not "us." (For an object-oriented ontological perspective, see the discussion of "spectral attunement" in Mickey, this volume.) With respect to cognition, in particular, this changing paradigm seems to support "the corporeal embeddedness of cognitive processes in the visceral dynamics of the brain, eye and skin, etc., and the con-figuration of human well-being with and through that of other living beings" (Whatmore 2002, 157, referencing the work of biologists Humberto Matuarana and Francisco Varela).

Multiplicity and decentering

In their editorial "Posthuman Geographies," Castree and Nash (2006) comment that "new scientific developments trouble the foundational figure of the human subject as distinct from other animal forms of life" (p. 501). Here, they are referring primarily to biotechnological developments that allow for interspecific transfers of such things as organs or genes. In light of the scientific material reviewed in this paper, however, Castree and Nash's statement takes on a new dimension. We humans are not only composed of and cannot exist without multiple other forms of life; we also share this multiplicity of being with other animals. In *The Posthuman*, Braidotti (2013) claims that posthumanist theory can "help us re-think the basic unit of reference for the human in the bio-genetic age known as 'Anthropocene,' the historical moment when the Human has become a geological force capable of affecting all life on this planet" (p. 5).

The foregoing discussion of gut bacteria gives another entry point into posthumanist theorizing of the current time period (and state of crisis) in which we find ourselves. Scholarship from decades ago (e.g., Deleuze and Guattari 1988; Haraway 1991; Latour 1993) challenges us to think of the material world in hybrid terms, as composed of assemblages and multiplicity, rather than fixed, pre-existing entities. For Deleuze and Guattari (1988), the concept of "multiplicity" was "created precisely in order to escape the abstract opposition between the multiple and the one, to escape dialectics, to succeed in conceiving the multiple in the pure state" (p. 32). Ontologically speaking, the multiple is of the first order, not something that arises by intentional or random combining of unitary forms. Further, it is not just that we are multiple. As Mary Midgley stated in *Utopias, Dolphins, and Computers* (1996), "[W]e are not self-contained and self-sufficient, either as a species or as individuals, but live naturally in deep mutual dependence" (quoted in Braidotti 2013, 77). For many of these scholars, the idea of multiplicity and interdependence was likely primarily conceived of in terms of exteriority and did not give full consideration to the literal assemblage of an organism through its (gut) microbiome. Braidotti gets closer to this when she discusses redefining the self, stating that "the subject is a transversal entity, fully immersed in and immanent to a network of non-human (animal, vegetable, viral) relations" (p. 193). However, just before this statement, Braidotti refers to "the collective nature and outward-bound direction of what we call the self" (p. 193). Our collective nature, as the assemblages in our guts demonstrate, is also "inward bound." What all of the foregoing—posthuman theory and increasing knowledge of the composition of bodies—points to is that the human body does not exist as such; what we call "the self" is not only an ideological construct, but also likely determined in many ways by the totality of beings that make up the human body, and that that most revered of human qualities in Western society—the human brain and its capacity for rational thought—may also be largely determined by entities that the (alleged) human does not really control. In a 2015 interview, the neuroscientist John Cryan theorized on the evolutionary relationship between "hosts'" brains and their gut microbes, suggesting that the microorganisms would have been able to evolve to shape the hosts' behavior to boost their own survival, saying that "happy people tend to be more social. And the more social we are, the more chances the microbes have to exchange and spread" (Kohn 2015). This multiplicity and entanglement of being should help us to decenter our long-held notions of the human and, further, ideas that our human agency is paramount to the shaping of what happens on our planet.

The scientific literature reviewed here illustrates not only that we humans are less human than we might have thought, but also that the changing global climate may impact all organisms in that most intimate of spaces—the gut. The study on the gut microbes of lizards points to negative impacts of warming temperatures, and the other studies indicate a closer relationship between the gut and the external environment than may have been previously thought. Temperatures, soils, crops—all will be affected by climate change, and these changes will find their way into bodies one way or another. This bodily impact will be shared all along the taxonomic spectrum and will, further, expand out across surrounding ecosystems.

A posthuman "climate"

Cary Wolfe suggests in *Animal Rites* (2003) that those looking to formulate posthuman ethical frameworks should not try to move beyond "the human" too quickly. "The human" (and all its attendant problems) is, after all, the framework through which we know the world. It is also not possible to leave such a durable concept and the institutions built around it completely behind. It is true that we cannot simply slough the skin through which we recognize ourselves and our world. Posthumanist scholarship at least implicitly recognizes the durability of humanism and its ideals. However, Braidotti also comments on the durability of life *as life* on this planet, referencing the "staggering, unexpected and relentlessly generative ways in which life . . . keeps on fighting back" (p. 195), although it is not clear whether Braidotti had microbial life foremost in mind with this statement. To be sure, earthly life does adapt and evolve. However, we can think back to Gould's opinions referenced earlier in this chapter. The forms of life we cannot see, and therefore tend to forget about (until they cause us a problem), are relentless in their persistence. What we currently call "the human" can't hold a candle to them. This microscopic life severely challenges both the "Anthropocene" and the "anthropo-scene."

How we view the world around us and our obligations to it is shaped by how we view ourselves. To get to different ethics, we must strive for a posthuman "climate." Nichols (this volume) asserts that posthumanist thought can further the development of "ecocentric" ethical frameworks which recognize the value of the entire biophysical world. In more general terms, posthumanism sees the disruption of the human/nature (or human/everything else) binary, the disidentification with "anthropo," and taking a nonessentialist perspective as having ethical import (Whatmore 2002; Lorimer 2009; Braidotti 2013). According to Braidotti (although with reference to techno-culture), trans-species connections are increasingly being recognized, and she advocates for a "deterritorialization" in order to "bypass metaphysics of substance . . . and the dialectics of otherness" (p. 71). Here, I suggest we do not necessarily need to refer to techno-culture for this deterritorialization. We can achieve the bypass Braidotti encourages by considering the gut microbiome. Substance becomes more fluid and our conceptions of otherness transformed and made less absolute when we recognize our organic bodily multiplicity. Therefore, perhaps this is a *re*-territorialization.

I agree with Braidotti's claim that posthumanism requires that our ethics be up to the challenge of the complex nature of the time in which we live. The two worldly contexts of this chapter—climate change and gut microbiomes—are also complex. The vastness and interconnectedness of climate change causes and effects make full understanding and prediction impossible. (See also the discussion of "hyperobjects" in Mickey, this volume.) Even at the apparently small scale of the animal gut, one encounters causes and effects similarly pushing the limits of comprehension. As Braidotti notes, the potential for widespread catastrophe caused by climate change links the fate of all earthly species, and I would add that the functioning and possible disruption of the space of the gut further connects all earthly animal life.

Following Braidotti, I contend that only by disrupting the idea/l of the human (self) can we take a non-anthropocentric view of ethics in the context of climate change (or other contexts as well). Despite wanting this disruption, I recognize that I have talked quite a bit about the human in this chapter. Unfortunately, in challenging a concept, institution, or perceived reality, we frequently have to reinscribe the object as part of that challenge. Or, as Wolfe states, "[T]he only way out is through" (p. 207). In the interest of "getting out," I want to reiterate that those most basic of bodily functions—eating, digesting, excreting—have always connected all animals, and now, with the possibility of climate change affecting those functions, we have an additional dimension of that connection. We are seeing more evidence that this part of the body not only challenges the discreteness of any species or individual, but also that which has been considered to be quintessentially human—the functioning of the brain and the ability to have a unitary self. Perhaps changing ontologies will change our literal and figurative climate to a more salutary posthuman one.

Notes

1 Posthumanist scholarship's roots connect it with other areas of critical scholarship such as deconstruction, feminist theory, postmodernism, and poststructuralism. Posthumanism began to be recognized as its own branch of scholarship roughly in the 1990s to the

early 2000s, as numerous scholars took on the project of questioning the precepts of the Western humanist intellectual tradition, which, in short, accepts "the human" as the (implicit or explicit) central figure of importance in any realm of inquiry. Badmington (2004) also notes limited instances of the term "posthuman" being used much earlier, as far back as the late 1800s and also in the 1970s.
2　This graph was originally included as a figure in Mann et al. (1999). Although the precise configuration of the graph and the modelling used to construct it became a subject of debate (see Le Page 2007 for a synopsis), the underlying findings have been upheld.
3　Although credit is now more widely noted to be shared with Alfred Russel Wallace, who, independently, arrived at substantially the same theory as did Darwin.

References

Agamben, G. *The Open: Man and Animal*. Stanford, CA: Stanford University Press, 2004.

Autism Speaks. *Grant Search: Defining the Underlying Biology of Gastrointestinal Dysfunction in Autism*. 2018. Accessed December 10, 2018. https://science.grants.autismspeaks. org/search/grants/defining-underlying-biologygastrointestinal-dysfunction-autism.

Badmington, N. "Mapping Posthumanism." *Environment and Planning A: Economy and Space* 36 (2004): 1344–51.

Bestian, E., J. Staffan, L. Zinger, L. Di Gesu, M. Richard, J. White, and J. Cote. "Climate Warming Reduces Gut Microbiota Diversity in a Vertebrate Ectotherm." *Nature Ecology & Evolution* 1, no. 161 (2017): 1–3.

Bolt, G. "Go with Your Gut." *Cascade: Magazine of the UO College of Arts & Sciences* (Winter 2015): 12–14.

Braidotti, R. *The Posthuman*. Malden, MA: Polity Press, 2013.

Braun, B. "Modalities of Posthumanism." *Environment and Planning A: Economy and Space* 36 (2004): 1352–55.

Braun, B. "Querying Posthumanisms." *Geoforum* 35 (2004): 269–73.

Butler, J. *Bodies That Matter: On the Discursive Limits of Sex*. New York: Routledge, 1993.

Castree, N. and C. Nash. "Editorial: Posthuman Geographies." *Social & Cultural Geography* 7, no. 4 (2006): 501–4.

Chakrabarty, D. "The Climate of History: Four Theses." *Critical Inquiry* 35 (2009): 197–222.

Darwin, C. *On the Origin of Species*. London: John Murray, 1859.

Deleuze, G. and F. Guattari. *A Thousand Plateaus: Capitalism and Schizophrenia*. London: Athlone Press, 1988.

Derrida, J. *The Animal That Therefore I Am*. Edited by M. Mallet and translated by D. Wills. New York: Fordham University Press, 2008.

Elliott, D. "In Autism, the Importance of the Gut." *The Atlantic*, June 10, 2013. Accessed December 10, 2018. www.theatlantic.com/health/archive/2013/06/in-autism-the-importance-of-thegut/276648/.

Foucault, M. *Power/Knowledge: Selected Interviews and Other Writings, 1972–1977*. Edited by C. Gordon. New York: Pantheon Books, 1980.

Fukuyama, F. *Our Posthuman Future: Consequences of the Biotechnology Revolution*. London: Profile Books, 2002.

Gershon, M.D. *The Second Brain: A Groundbreaking New Understanding of Nervous Disorders of the Stomach and Intestine*. New York: HarperCollins Publishers, 1998.

Gould, S.J. "Planet of the Bacteria." *The Washington Post*, November 13, 1996, H01.

Graham, E.L. *Representations of the Post/Human: Monsters, Aliens and Others in Popular Culture*. Manchester, UK: Manchester University Press, 2002.

Guggenheim, D. *An Inconvenient Truth*. Los Angeles: Lawrence Bender Productions and Participant Productions, 2006.

Gurry, T., HST Microbiome Consortium, S.M. Gibbons, L.T.T. Nguyen, S.M. Kearney, A. Ananthakrishnan, X. Jiang, C. Duvallet, Z. Kassam, and E.J. Alm. "Predictability and Persistence of Prebiotic Dietary Supplementation in a Healthy Human Cohort." *Scientific Reports* 8 (2018): 1–13.

Hadhazy, A. "Think Twice: How the Gut's 'Second Brain' Influences Mood and Well-Being." *Scientific American*, February 12, 2010. Accessed December 10, 2018. www.scientificamerican.com/article/gut-second-brain/.

Haraway, D. *Simians, Cyborgs, and Women: The Reinvention of Nature*. London: Routledge, 1991.

IPCC. *Climate Change 2014: Synthesis Report. Contribution of Working Groups I, II and III to the Fifth Assessment Report of the Intergovernmental Panel on Climate Change*. Edited by Core Writing Team, R.K. Pachauri and L.A. Meyer. Geneva, Switzerland: IPCC, 2014.

Jandhyala, S.M., R. Talukdar, C. Subramanyam, H. Vuyyuru, M. Sasikala, and D.N. Reddy. "Role of the Normal Gut Microbiota." *World Journal of Gastroenterology* 21, no. 29 (2015): 8787–803.

Johnston, C.L. "Taxonomy." In *Humans and Animals: A Geography of Coexistence*, edited by J. Urbanik and C.L. Johnston, 320–22. Santa Barbara, CA: ABC-CLIO, 2017.

Kohn, D. "When Gut Bacteria Change Brain Function." *The Atlantic*, June 24, 2015. Accessed December 10, 2018. www.theatlantic.com/health/archive/2015/06/gut-bacteria-on-thebrain/395918/.

Latour, B. *We Have Never Been Modern*. Cambridge: Harvard University Press, 1993.

Le Page, M. "Climate Myths: The 'Hockey Stick' Graph Has Been Proven Wrong." *New Scientist*, May 16, 2007. Accessed January 31, 2019. www.newscientist.com/article/dn11646-climate-myths-the-hockey-stick-graph-has-been-proven-wrong/.

Li, Q., Y. Han, A.B.C. Dy, and R.J. Hagerman. "The Gut Microbiota and Autism Spectrum Disorders." *Frontiers in Cellular Neuroscience* 11, no. 120 (2017): 1–14.

Lorimer, J. "Posthumanism/Posthumanistic Geographies." In *International Encyclopedia of Human Geography*, edited by R. Kitchin and N. Thrift, 344–54. Oxford: Elsevier, 2009.

Mann, M.E. "Little Ice Age and Medieval Climatic Optimum." In *Encyclopedia of Global Environmental Change*, edited by M.C. McCracken and J.S. Perry, vol. 1., 504–9, 514–16. Chichester, UK: John T. Wiley & Sons, 2002.

Mann, M.E., R.S. Bradley, and M.K. Hughes. "Northern Hemisphere Temperatures During the Past: Inferences, Uncertainties, and Limitations." *Geophysical Research Letters* 26, no. 6 (1999): 759–62.

Midgley, M. *Utopias, Dolphins and Computers: Problems of Philosophical Plumbing*. New York: Routledge, 1996.

Mitrovica, J.X., C.C. Hay, E. Morrow, R.E. Kopp, M. Dumberry, and S. Stanley. "Reconciling Past Changes in Earth's Rotation with 20th Century Global Sea-Level Rise: Resolving Munk's Enigma." *Science Advances* 1, no. 11 (2015): 1–6.

Moyer, M.W. "Gut Bacteria May Play a Role in Autism." *Scientific American*, September 1, 2014. Accessed October 5, 2018. www.scientificamerican.com/article/gut-bacteria-may-play-a-role-in-autism/.

Peddada, S. "Seasonal Change in the Gut." *Science* 357, no. 6353 (2017): 754–55.

Plumwood, V. *Feminism and the Master of Nature*. London: Routledge, 1993.

Said, E. *Humanism and Democratic Criticism*. New York: Columbia University Press, 2004.

Schiebinger, L. *Nature's Body: Gender in the Making of Modern Science*. New Brunswick, NJ: Rutgers University Press, 2008.

Smits, S.A., J. Leach, E.D. Sonnenburg, C.G. Gonzalez, J.S. Lichtman, G. Reid, R. Knight, A. Manjurano, J. Changalucha, J.E. Elias, M.G. Dominguez-Bello, and J.L. Sonnenburg. "Seasonal Cycling in the Gut Microbiome of the Hadza Hunter-Gatherers of Tanzania." *Science* 357, no. 6353 (2017): 802–6.

Tasnim, N., N. Abulizi, J. Pither, M.M. Hart, and D.L. Gibson. "Linking the Gut Microbial Ecosystem with the Environment: Does Gut Health Depend on Where We Live?" *Frontiers in Microbiology* 8 (2017): 1–8.

Thursby, E. and N. Juge. "Introduction to the Human Gut Microbiota." *Biochemical Journal* 474 (2017): 1823–36.

Whatmore, S. *Hybrid Geographies: Natures Cultures Spaces*. Thousand Oaks, CA: Sage Publications, Inc., 2002.

Wolfe, C. *Animal Rites: American Culture, the Discourse of Species, and Posthumanist Theory*. Chicago: The University of Chicago Press, 2003.

14 Wonderland Earth in the Anthropocene epoch

Holmes Rolston III

> Wonders are many, and none is more wonderful than man.
>
> – Sophocles, *Antigone*

Wonderland planet

Earth is, by all accounts, a wonderland planet. Let's take that first from rocket science. Viewing Earth from space, the astronaut Michael Collins recalled being earthstruck: "Earth is to be treasured and nurtured, something precious that must endure" (Collins 1980, 6). No one contests that: scientists, philosophers, politicians, economists, theologians, business executives, farmers, housewives, ordinary people. Whole Earth photographs from space are as widely viewed as any in human history and, in pensive moments, invariably give viewers pause to wonder at their stunning home planet.

In the cosmos, remarkable features produce billions of galaxies, with stars generating elements suitable for forming planets. Life is so far known only on planet Earth, where over billions of years, there has been an explosion of life, moving through several billion species, reaching humans in an evolutionary process. By widespread scientific accounts, humans result from a cosmic "anthropic" principle.

In the last half century, scientists have found dramatic interrelationships between astronomical and atomic scales that connect to make the universe "user-friendly." Astronomical phenomena, such as the formation of galaxies, stars, and planets, depend critically on the microphysical phenomena. In turn, those mid-range scales, where the known complexity mostly lies, depend on the interacting microscopic and astronomical ranges. The stars are the furnaces in which all but the very lightest elements are forged. The stars run their courses, and many explode as supernovae to disperse their matter throughout space. Such matter is condensed as planets, and life evolves out of such elements.

If the scale of the universe were much reduced, there would not have been enough time for elements to form. If the expansion rate of the universe had been a little faster or slower, then the universe would already have recollapsed or the galaxies and stars would not have formed. How fast the universe is expanding depends on the value of what physicists call the "cosmological constant." They symbolize this with the Greek letter λ. This constant is quite small, nearly zero but not zero. Martin Rees, leading British astronomer at Cambridge University, reflects,

> Fortunately for us (and very surprisingly to theorists) λ is very small. Otherwise its effect would have stopped galaxies and stars from forming and cosmic evolution would have been stifled before it could even begin. . . . The cosmic number λ – describing the weakest force in nature, as well as the most mysterious – seems to control the universe's expansion and its eventual fate.
>
> (Rees 2000, 3, 98–99)

John Barrow, physicist and mathematician at Cambridge, looks out at the universe: "Many of its most striking features – its vast size and huge age, the loneliness and darkness of space – are all necessary conditions for there to be intelligent observers like ourselves" (Barrow 2002, 113).

Four fundamental forces hold the world together: the strong nuclear force, the weak force, electromagnetism, gravitation. Change slightly the strengths of any of those four forces, change critical particle masses and charges, and the stars would burn too quickly or too slowly, or atoms and molecules (including water, carbon, and oxygen) or amino acids (building blocks of life) would not form or remain stable. We have discovered that what seem to be widely varied facts really cannot vary widely, indeed, that many of them can hardly vary at all, and have the universe develop the matter, life, and mind it has generated. Roger Penrose, physicist and mathematician at Oxford, is impressed by "the extraordinary degree of precision or 'fine-tuning' for a big bang of the nature that we appear to observe." He concludes that ours is "an extraordinarily special Big Bang" (Penrose 2005, 726, 762). That big bang is now thought to have been an explosion resulting from a fluctuation in a quantum vacuum.

Paul Davies, a cosmologist formerly at Cambridge, now at the Arizona State University, claims that we hit "the cosmic jackpot," a universe "just right for life" (Davies 2007). Max Tegmark, cosmologist at MIT, phrases this more technically: "virtually no physical parameters can be changed by large amounts without causing radical qualitative changes to the physical world. In other words, the 'island' in parameter space that supports human life appears to be quite small" (Tegmark 1998, 6).

Where once there were no species on Earth, there are today five to ten million. Prokaryotes dominated the living world more than three billion years ago; there later appeared eukaryotes, with their well-organized nucleus and cytoplasmic organelles. Single-celled eukaryotes evolved into multi-celled plants and animals with highly specialized organ systems. First, there were cold-blooded animals at the mercy of climate and later warm-blooded animals with more energetic metabolisms. From small brains emerge large central nervous systems. Although biologists continue to debate "progress" in natural history, we need to put some kind of an arrow on evolutionary time.

The life story is different, because in biology, unlike physics, chemistry, geomorphology, or astronomy, something can be learned. Genes are cybernetic units of inheritance, capable of discovering and storing life information and elaborating it. Such functional agency is a novel wonder on Earth. The novelty is that matter-energy enters into information states. With its genetic coding, an organism is "informed" about how to make a way through the world, how to cope in its niche. Past achievements are recapitulated in the present, with variations; these results get tested today and then folded into the future. Random mutation figures into a larger generative process; species generate and test new possibilities. The challenge is to get as much versatility coupled with as much stability as is possible. This requires optimizing twin maxima, keeping past knowledge while exploring the nearby search space for better adaptation.

Organisms compete, struggling to hold a place against other lives. To be alive is to have problems. Survival is the name of the game. Yet in a more inclusive perspective, the idea of adapted fit also requires a niche, a place to be, and includes a life support system. An ecology is a home. The currents of life flow in the interplay of environmental conductance and environmental resistance. An environment that was entirely hostile would slay all; life could never have appeared within it. An environment that was entirely irenic would stagnate life. The vital natural process is of conflict and resolution. The organism is tested for how much information it can contribute to the next generation. Survival of the fittest turns out to be survival of the senders.

The strange wonder now is that the cosmic start-up seems a setup for life, necessary but not sufficient. Yet such life is rare in the universe and exploding on Earth. Life starts up and then smarts up. Scientists have found other planets, currently over a thousand of them. But those on which life seems possible (in the range of liquid water and adequate energy) are rare among them. Wonderland Earth is necessary and sufficient for life. "It appears that Earth got it just right," concluded Peter D. Ward and Donald Brownlee, a geologist and an astronomer, celebrating "Rare Earth" and noting why complex life will be rare in the universe (Ward and Brownlee 2000, 265). Lewis Thomas, a famous biologist, celebrates how Earth is "the only exuberant thing in this part of the cosmos" (Thomas 1975, 145). A good planet is hard to find.

Humans—the wonder of wonders

On Earth, humans are, by all accounts, the most complex and startling species. Humans are endowed with a genetic heritage producing the human mind, by far the most complex thing known, of virtually infinite complexity, capable of semantic and symbolic speech. With such mind, they generate cumulative transmissible cultures, elaborating high orders of rational and emotional thought in science, philosophy, ethics, and religious faith. Humans alone ask who they are, where they are, and what they ought to do. That humans evolved out of fossil stardust, creatively generated out of a fluctuation in a quantum vacuum, is quite a miracle.

The explosive growth of the human brain, sponsoring the cognitively spectacular human mind is the principal wonder on Earth. Edward O. Wilson, Harvard University, remarks, "No organ in the history of life has grown faster" (Wilson 1978, 87). Steve Dorus and his team of neurogeneticists, University of Chicago, conclude, "Human evolution is characterized by a dramatic increase in brain size and complexity" (Dorus et al. 2004, 1027). J. Craig Venter and over 200 co-author geneticists call the human brain "a massive singularity" (Venter et al. 2001, 1347).

Michael Gazzaniga, prominent neuroscientist, University of California, Santa Barbara, speaks of "the explosion in human brain size":

> We are hugely different. While most of our genes and brain architecture are held in common with animals, there are always differences to be found. And while we can use lathes to mill fine jewelry, and chimps can use stones to crack open nuts, the differences are light years apart. . . . We humans are special.
>
> (Gazzaniga 2008, 1–3, 13)

The human brain is not just a scaled-up version of a chimpanzee brain. Humans are remarkable in their capacities to process thoughts, ideas, and symbolic abstractions figured into interpretive gestalts with which the world is understood and life is oriented. This higher consciousness is a constitutive dimension of humans and is absent in all other species. The key threshold is the capacity to pass ideas from mind to mind. There is no clear evidence that chimpanzees attribute mental states to others.

Chimps have little or no "theory of mind"; they do not know other minds are there with whom they might communicate, to learn what they know. Or, if you prefer to say that one chimp can know what another knows, chimps have a theory of immediate mind (one chimp sees that another chimp knows where those bananas are); humans have a theory of the ideational mind (one human teaches another the Pythagorean theorem). Humans have ideational uniqueness. Their cultural transmission makes it possible for an individual to inherit the discoveries of thousands of others before him, discoveries that the individual could not make in a single lifetime.

Although chimpanzees collaborate to hunt or get food, Michael Tomasello and his colleagues at the Max Planck Institute for Evolutionary Anthropology, Leipzig, conclude,

> It may be said with confidence that chimpanzees do not engage in collaborative learning. . . . They do not conceive of others as reflective agents – they do not mentally simulate the perspective of another person or chimpanzee simulating their perspective. . . . There is no known evidence that chimpanzees, whatever their background and training, are capable of thinking of other interactants reflectively.
>
> (Tomasello et al. 1993, 504–5)

Daniel Povinelli, biologist at the University of Louisiana, Lafayette, concludes, "Humans have a whole system that we call theory of mind that chimps don't have" (Povinelli, quoted in Pennisi 1999, 2076).

Some trans-genetic threshold seems to have been crossed. The human brain is of such complexity that descriptive numbers are astronomical and difficult to fathom. A typical estimate is 10^{12} neurons, each with several thousand synapses (possibly tens of thousands). Each neuron can "talk" to many others. The postsynaptic membrane contains over a thousand different proteins in the signal receiving surface. "The most molecularly complex structure known [in the human body] is the postsynaptic side of the synapse," according to Seth Grant, a neuroscientist at the University of Edinburgh (quoted in Pennisi 2006). Over a hundred of these proteins were co-opted from previous, non-neural uses; by far, the most of them evolved during brain evolution. This is nature's nanotechnology.

This nanophysiology is integrated into a dendritic network structured at multiple hierarchical levels. This network, formed and re-formed, makes possible virtually endless mental activity. Much, even most, of what goes on in our brains is below the level of conscious awareness, of course, but humans can bring to critical focus novel cognitive capacities. The result is a mental combinatorial explosion. The human brain is capable of forming thoughts numbering something in the range of $10^{70,000,000,000}$ thoughts—a number that dwarfs the number of atoms in the visible universe (10^{80}) (Flanagan 1992, 37; Holderness 2001). On a cosmic scale, humans are minuscule atoms, but on a complexity scale, humans have "hyperimmense" possibilities in mental complexity (Scott 1995, 81). In our 150 pounds of protoplasm, in our three-pound brain is more operational organization than in the whole of the Andromeda galaxy.

Genes make the kind of human brains possible that facilitate an open mind. But when that happens, these processes can also work the other way around. What began as a "bottom-up" process becomes a "top-down" process. In "top down" causation, an emergent phenomenon re-shapes and controls its precedents, as contrasted with "bottom up" causation, in which precedent, simpler causes are fully determinative of more complex outcomes. Minds employ and reshape their brains to facilitate their chosen ideologies and lifestyles. Our ideas and our practices configure and re-configure our own sponsoring brain structures.

The linguistic ideational uniqueness involves complex use of symbols. Ian Tattersall, archaeologist at the American Museum of Natural History, New York, concludes, "We human beings are indeed mysterious animals. We are linked to the living world, but we are sharply distinguished by our cognitive powers, and much of our behavior is conditioned by abstract and symbolic concerns" (Tattersall 1998, 3). Similarly, Richard Potts, Smithsonian Institution's National Museum of Natural History, concludes, "All the odd elaborations of human life, socially and individually, including the heights of imagination, the depths of depravity, moral abstraction, and a sense of God, depend on this *symbolic coding of the nonvisible* (Potts 2004, 263). So humans alone produce both theory and practice, in science and mathematics, in ethics and politics, and in religious faith.

Humans find themselves uniquely emplaced on a unique planet—in their world cognitively and critically as no other species is. Our bodily incarnation embeds us in this biospheric community; we are Earthlings. Our mental genius enables us to rise to transcending overview. So we can conclude that on this wonderland Earth, we *Homo sapiens* are the wonder of wonders. We can conclude that we are genius on top. Next, we have to wonder what that can mean—how can and ought we to be on top?

Wondering about Anthropocene humans?

By recent accounts, human dominance is so extensive that Earth has entered a new age, the Anthropocene epoch (Crutzen 2006a). The mental activity of humans reshaping their agentive capacities has in recent centuries produced technological development giving humans vast powers for transforming their planet through agriculture, industry, and technology. This has so dramatically escalated that we have entered the first century in the 45 million centuries of life on Earth in which one species can aspire to manage the planet's future.

What is the empirical evidence? Anthropocene enthusiasts say, "Just look, anywhere, everywhere. Human-dominated ecosystems cover more of Earth's land surface than do wild ecosystems" (McCloskey and Spalding 1989; Foley et al. 2005). Human agriculture, construction, and mining move more Earth than do the natural processes of rock uplift and erosion. Humans are now the most important geomorphic agent on the planet's surface (Wilkinson and McElroy 2007). "Human activities have become so pervasive and profound that they rival the great forces of Nature and are pushing the Earth into planetary *terra incognita*" (Steffen et al. 2007, 614). Geologists need stratigraphic evidence. The International Commission on Stratigraphy has a working group that has recommended Anthropocene as a geological unit (Waters et al. 2016).

Beyond the geology, "Anthropocene" has become an "elevator word" and put to use philosophically. *The Economist* has a cover story: "Welcome to the Anthropocene." "A Man-Made World." "The challenge of the Anthropocene is to use human ingenuity to set things up so that the planet can accomplish its 21st century task." They foresee "10 billion reasonably rich people" on a geoengineered, genetically synthetic Earth, re-built with humans in center focus (*Economist* 2011, 11, 81). Capitalist markets and the media feature increased fulfilling and expanding of human wants. The Anthropocene is "humanity's defining moment," according to the American Geosciences Institute (Seielstad 2012). "Humans are the ultimate ecosystem engineers" (Ellis and Ramankutty 2009). We are "the God species" (Lynas 2011).

Since Galileo, Earth seemed a minor planet, lost in the stars. Since Darwin, humans have come late and last on this lonely planet. Today, on our home planet at least, we are putting these once de-centered humans back at the center. We have entered the era of the imperial human domain. "What we call 'saving the Earth' will, in practice, require creating and re-creating it again and again for as long as humans inhabit it" (Shellenberger and Nordhaus 2011, 61). Humans are now "too big for nature." "Let us embrace the challenge to gain mastery over human engagement with the earth" (Ellis 2015). Enter the designer world.

This is illustrated in how human changes to the planet are producing global warming. Humans do not need ever again to face ice ages, as they did in the Pleistocene. Allen Thompson, an environmental philosopher, with a "radical hope for living well in a warmer world," urges us to find a significantly "diminished place for valuing naturalness," replacing it with a new kind of "environmental goodness . . . distinct from nature's autonomy" (Thompson 2010, 43, 56). Erle Ellis, in what he calls the "Planet of No Return: Human Resilience on an Artificial Earth," celebrates "the beginning of a new geological epoch ripe with human-directed opportunity" (Ellis 2011, 44).

This is forcing humans to re-evaluate their role on the Earth in the future of their planet.

Yes, on Earth, humans are the wonder of wonders. But now we begin wonder about these Anthropocene humans—in a different sense of "wonder"—that of doubt and uncertainty, not of astonishment and marvel.

Managed planet and end of nature?

Enthusiasts for increasing human powers advocate that humans can and ought to manage their planet in their self-interest, engineering Earth resourcefully for increasing human benefits, bringing about the end of nature. Humans are now the most important geomorphic agent on the planet's surface. We should embrace the Anthropocene. We should use human ingenuity for an ever-escalating technology, ever-increasing human domination of the landscape, perpetual enlargement of the bounds of the human empire. In this mood, the Anthropocene enthusiasts are gung-ho for change.

The editors of a *Scientific American* special issue, *Managing Planet Earth*, ask, "What kind of planet do we want? What kind of planet can we get?" (Clark 1989). Find ways to redistribute rainfall, stop hurricanes and tsunamis, prevent earthquakes, redirect ocean currents, fertilize marine fisheries, manage sea-levels, alter landscapes for better food production, and generally make nature more user-friendly. Edward Yoxen urges,

> The living world can now be viewed as a vast organic Lego kit inviting combination, hybridisation, and continual rebuilding. Life is manipulability. . . .
> Thus our image of nature is coming more and more to emphasise human intervention through a process of design.
>
> (Yoxen 1983, 15)

"The biosphere itself, at levels from the genetic to the landscape, is increasingly a human product (Allenby 2000, 11). We live in "anthropogenic biomes" (Ellis and Ramankutty 2008).

Geoengineering is "the intentional large-scale manipulation of the environment" (Keith 2000, 245). Paul Crutzen, the climate scientist who has dramatized the term "Anthropocene," argues that geoengineering "should be explored," given the dismal prospects of any other solution (Crutzen 2006b, 212). "The time has come to take it seriously. Geoengineering could provide a useful defense for the planet – an emergency shield that could be deployed if surprisingly nasty climatic shifts put vital ecosystems and billions of people at risk" (Victor et al. 2009, 66; Launder and Thompson 2010). There are several possibilities: Launch reflective particles into the upper atmosphere or aerosols or a cloud of thin refracting disks or reflective balloons, thereby cooling the Earth, as volcanic eruptions have done in the past. Or fertilize the ocean so as to increase plankton, which absorb more carbon. Or spray fine ocean water mist into the clouds to make them brighter, reflecting more sunlight. There are technological challenges to all these proposals. "Such schemes are fraught with uncertainties and potential negative effects" (Blackstock and Long 2010).

None of this sounds like humans intelligently re-engineering the planet on which they find themselves, rationally planning for an Anthropocene age. It sounds more like panic on a planet that the engineers are realizing that they have messed up, in ways they find almost beyond their control. Humans are smarter than ever, so smart that we are faced with overshoot (Dilworth 2010). Our power to make changes exceeds our power to predict the results, exceeds our power to control even those adverse results we may foresee.

There is concern about ending nature on Earth. "We live at the end of nature, the moment when the essential character of the world . . . is suddenly changing." Bill McKibben worries that already "we live in a postnatural world," in "a world that is of our own making." "There's no such thing as nature anymore" (McKibben 1989, 60, 85, 89, 175). There is only the built environment. Michael Soulé faces this prospect: "The term natural will disappear from our working vocabulary. The term is already meaningless in most parts of the world because anthropogenic [activities] have been changing the physical and biological environment for centuries, if not millennia" (Soulé 1989, 301). We are at "the end of the wild" (Meyer 2006). We are "living through the end of nature" (Wapner 2010). Nature is over.

"Human beings are at the centre of concerns. . . " So the *Rio Declaration* begins, formulated at the United Nations Conference on Environment and Development and signed by almost every nation on Earth (UNCED 1992). This was once to be called the *Earth Charter*, but the developing nations were more interested in asserting their rights to develop and only secondarily in saving the Earth. The Rio claim is, in many respects, quite true. The human species is causing all the concern. The problem is to get people into "a healthy and productive life in harmony with nature" (UNCED 1992).

Anthropocene arrogance

Critics wonder about Anthropocene arrogance. A planet we manage, or attempt to manage, only to secure more and more profits and commodities for ourselves reveals an exploitive frame of mind. We shape our worldviews, and then our worldviews shape us. We fear that humans have become Earth's global consumer, Earth's juggernaut predator. The Anthropocene is colonialism resurrected in super global form. An overweening pride, *hubris*, is by many classical accounts the original human sin. "You shall be as gods" (Genesis 3).

What we must push for, according to the Royal Society of London, is "sustainable intensification" of reaping the benefits of exploiting the Earth (*Royal Society* 2009). Would not the world's oldest scientific society be as well advised to ask about protecting ancient and ongoing biodiversity, about how we might shrink our footprint, whether treading softly is wiser than ever intensifying our imperial exploitation? If we are to fix the problem in the right place, we must learn to manage ourselves as much as the planet.

David Biello, *Scientific American*'s energy and environment editor, exclaims,

> The stakes could not be higher. . . . What we stand to gain is nothing less than an enduring civilization and a firmer understanding of our planet and ourselves. We have arrived at a new geologic epoch of our own making. . . . I argue the goal must be to make an enduring Anthropocene, an epoch that, in geologic and civilizational terms, stretches into an era. . . . This is not the end of the world. This is just the end of the world as we have known it.
>
> (Biello 2016, 7–8)

Anthropocentric enthusiasts make the claim that such power is to be welcomed ethically. For all of human history, we have been pushing back limits. Especially in the West, we have lived with a deep-seated belief that life will get better, that one should hope for abundance and work toward obtaining it. Economists call such behavior "rational." Ethicists can agree: We ought to maximize human satisfactions, the abundant life, with more and more of the goods and services that people want. We have a right to self-development, to self-realization. Such growth, always desirable, is now increasingly possible.

Anthropocene enthusiasts may here deny arrogance. Just the other way around. They take the moral high ground: Classical conservation has been "socially unjust" (Kareiva and Marvier 2012, 965). "Protecting nature that is dynamic and resilient, that is in our midst rather than far away, and that sustains human communities – these are the ways forward now. Otherwise, conservation will fail, clinging to its old myths." "Instead of pursuing the protection of biodiversity for biodiversity's sake, a new conservation should seek to enhance those natural systems that benefit the widest number of people, especially the poor" (Kareiva et al. 2011, 36–37).

The dream of living in harmony with nature is bygone. There is a more promising ambition: audacious humans manage their brave new world. Nature has been operating on the planet for five billion years. Human culture has been operating alongside and dependent on nature for something in the range of 40,000 to 100,000 years. Now the exuberant Anthropocene architects wish to displace globally systemic nature and radically shape the future as no generation before has had either the capacity or aspiration to do. And this will be a blessing in a more humane, equitable world.

Critics worry that, though the intentions sound high, they have an immoral trailer. "Forward for me and my kind!" "Save nature for people, not from people." That could be as much the problem as the answer. The subtext seems to be the "old myths" that wildlife or ecosystems or biodiversity or evolutionary creative genesis have goods of their own, intrinsic value worth protecting. Essentially this puts us as the first, if not the only, location of moral relevance. Justice is just-us. This is the Anthropocene, and too bad for the non-anthropic. Anthropocene proponents are concerned to get people fed, even if doing so drives tigers and butterflies into extinction.

Kareiva and Marvier urge us to shift "from a focus almost exclusively on biodiversity" to more attention to

> human well-being. . . . Conservation is fundamentally an expression of human values. . . . Today we need a more integrative approach in which the centrality of humans is recognized. . . . We do not wish to undermine the ethical motivations for conservation action. We argue that nature also merits conservation for very practical and more self-centered reasons concerning what nature and healthy ecosystems provide to humanity.
>
> (Kareiva and Marvier 2012, 963–65)

Despite the caveat, ethical concern for non-humans is soon undermined. We may be told that once-abundant species can vanish with no ill effects on humans—the bison, the chestnut, the passenger pigeon, the dodo, the tigers and butterflies. Putting ourselves first makes it difficult to appreciate the other-than-human.

Rebuilding the planet with humans at the center, or even protecting ecosystem services so long as these benefit us, no longer sounds like the high moral ground. Nature is of value only if and so far as it supports human enterprises. This puts the whole planet in the service of only one species—an unnatural condition. If our concern is for the poor in this new humanist excellence, then why not emphasize environmental justice, more equitable distribution of wealth between rich and poor on developed lands, rather than diminishing wild nature to benefit the poor?

On future Earth, it is hard to imagine a world without ongoing development—without engines and gears, without electricity, without cars, cell phones, computers. We expect ever-escalating high technology in our service. Self-fulfilling desires intoxicate us; we grow addicted to them. In the Anthropocene, we might indeed get more and more of what we want. But this might lead us to accept an environment increasingly toxic and degraded by global warming. This might lead our children not to notice their hotter, less diverse, less stable environment. We, our children, our children's children will never know our highest flourishing, dumbed down by our ever more assertive self-interests. "Quite possibly, then, this era, which so congratulates itself on its self-awareness, will come to be known as the time of the Great Derangement" (Ghosh 2016, 11).

Here is what Anthropocene proponents need first to confront. A massive *Millennium Ecosystem Assessment*, sponsored by the United Nations, involving over 1,300 experts from almost 100 nations, begins, "At the heart of this assessment is a stark warning. Human activity is putting such strain on the natural functions of Earth that the ability of the planet's ecosystems to sustain future generations can no longer be taken for granted" (*Millennium Ecosystem Assessment* 2005, 5). Encouraging a new Anthropocene epoch with ever-increasing human desires seems a deranged policy, far more likely to increase this strain than to reduce it. Anthropocene managers are unlikely to address harmful results, possible or probable, distant from themselves in time and space.

The geoengineers will find that their engineering is not just a technical problem; they have to consider the social contexts in which they launch their gigantic projects, the welfare and risks of those they seek to save, the (in)justice of geoengineering that spreads benefits and costs inequitably, the governance of geoengineering (Parson and Keith 2013). Engineers are no better equipped to deal with transdisciplinary systems problems than are the politicians. Or with the ethical problems. They may find a majority of Earth's residents wondering, Is our only relationship to nature one of engineering it for the better?

Now Allen Thompson, joined by Jeremy Bendik-Keymer, backs off, more inclined to work with, rather than revise, the basic processes in ecosystems. "Far from the current rush toward geoengineering, this kind of response would exhibit the virtue of humility" (Thompson and Bendik-Keymer 2012, 15). David Biello too worries about reckless Anthropocene over-management. "This is about managing change, adapting to it, and increasing the resilience of our civilization at the same time as we make more room for our fellow travelers on this life bearing spaceship" (Biello 2016, 7). We do not want Earth transformed into an artifact.

Several billion years' worth of creative toil, several million species of teeming life, have been handed over to the care of this late-coming species in which mind has flowered and morals have emerged. Ought not those of this sole moral species do something less self-interested than count all the produce of an evolutionary ecosystem resources to be valued and re-engineered only for the benefits they bring? Such an attitude hardly seems biologically informed, much less ethically adequate. Its logic is too provincial for moral humanity.

Wonderful humans incarnate on wonderland Earth

Humans coinhabit Earth with five to ten million other species, and we and they depend on surrounding biotic communities. There are multiple dimensions of naturalness, on both public and private lands. George Peterken, British ecologist, has an eight-point scale (Peterken 1996). Even on long-settled landscapes, there can be natural woodlands, treasured by owners over centuries. There may be native woodlands, often with quite old trees, secondary woodlands with trees fifty to a hundred years old, recently restored woodlands, wetlands, moors, hedgerows, mountains, such as the Alps or the Scottish Cairngorms. Gregory Aplet, a US forest ecologist, distinguishes 12 landscape zones, placed on axes of human "controlled" to autonomously "self-willed" and "pristine" to "novel." Rather than seeking to press onward toward totally managed Earth, why not claim that there are and ought to be various degrees of the preservation-conservation-Anthropocene spectrum?

Zoning the landscape, how much human management do we apply where? Which are urban lands? Which are working landscapes, rural or dedicated to multiple use? This "right-sizing" policy question seems to demand a more specific answer than we actually need to give, if we are concerned with sizing the human presence on future Earth. Wilderness is the most endangered landscape, the least-sized, the one in shortest supply. Save all you can. Right-size agricultural landscapes not by re-engineering weather, climate, and soil geologies but by right-sizing human populations at levels suitable as adapted fits in their supporting rural communities with ecosystem services. Technology can overcome some constraints (fossil fuels, nitrogen fertilizers) but only within ecosystem constraints (global warming, nitrogen-polluted waters). Right-size cities by keeping them sustainable on their supporting agricultures and ecosystems. Right-size humans by keeping them at home on their planet.

Rocket scientists, loving their marvelous, high-tech machines, are still concerned to celebrate our organic, vital planet. Viewing Earthrise from the moon, the astronaut Edgar Mitchell was entranced:

> Suddenly from behind the rim of the moon, in long, slow-motion moments of immense majesty, there emerges a sparkling blue and white jewel, a light, delicate sky-blue sphere laced with slowly swirling veils of white, rising gradually like a small pearl in a thick sea of black mystery. It takes more than a moment to fully realize this is Earth . . . home.
>
> (Mitchell, quoted in Kelley 1988, at photographs 42–45)

Humans are most wonderful, full of wonder, wonder-full when caring for this small, immensely majestic, precious pearl in the mystery of deep space. We do not want a de-natured life on a de-natured planet.

Our best hope is for a tapestry of cultural and natural values, not a trajectory even further into the Anthropocene. Keep nature in symbiosis with humans. Keep the urban, rural, and wild. Our future ought to be the Semi-Anthropocene, kept basically natural—with the natural basics—and entered carefully—full of cares for both humans and nature on this marvelous home planet. Cherish wonderful humans incarnate on wonderland Earth.

References

Allenby, B. "Earth Systems and Engineering and Management." *IEEE Technology and Society Magazine* 19, no. 4 (2000): 10–24.

Aplet, Gregory H. "On the Nature of Wildness: Exploring What Wilderness Really Protects." *University of Denver Law Review* 76 (1999): 347–67.

Barrow, John D. *The Constants of Nature.* New York: Pantheon Books, 2002.

Biello, David. *The Unnatural World: The Race to Remake Civilization in Earth's Newest Age.* New York: Simon and Schuster, Scribner, 2016.

Blackstock, J. J. and J. C. S. Long. "The Politics of Geoengineering." *Science* 327 (January 29, 2010): 527.

Clark, W. C. "Managing Planet Earth." *Scientific American* 261, no. 3 (September 1989): 46–54.

Collins, Michael. "Foreword." In *Our Universe*, edited by Roy A. Gallant. Washington, DC: National Geographic Society, 1980.

Crutzen, Paul J. "Albedo Enhancement by Stratospheric Sulfur Injections: A Contribution to Resolve a Policy Dilemma?" *Climatic Change* 77 (2006): 211–19.

Crutzen, Paul J. "The 'Anthropocene'." In *Earth System Science in the Anthropocene*, edited by Eckart Ehlers and Thomas Kraft, 13–18. Berlin: Springer, 2006.

Davies, Paul. *Cosmic Jackpot: Why Our Universe Is Just Right for Life*. Boston: Houghton Mifflin, 2007.

Dilworth, Craig. *Too Smart for Our Own Good: The Ecological Predicament of Humankind*. Cambridge: Cambridge University Press, 2010.

Dorus, Steve, et al. "Accelerated Evolution of Nervous System Genes in the Origin of Homo sapiens." *Cell* 119 (2004): 1027–40.

The Economist. "Welcome to the Anthropocene." 399, no. 8735 (2011).

Ellis, Erle. "The Planet of No Return." *Breakthrough Journal* 2 (Fall 2011): 39–44. http://thebreakthrough.org/index.php/journal/past-issues/issue-2/the-planet-of-no-return.

Ellis, Erle. "Too Big for Nature." In *After Preservation: Saving American Nature in the Age of Humans*, edited by Ben A. Minteer and Stephen J. Pyne, 24–31. Chicago: University of Chicago Press, 2015.

Ellis, Erle and Navin Ramankutty. "Anthropogenic Biomes." *Encyclopedia of Earth*. 2009. http://ecotope.org/people/ellis/papers/ellis_eoe_anthromes_2007.pdf.

Ellis, Erle and Navin Ramankutty. "Putting People in the Map: Anthropogenic Biomes of the World." *Frontiers in Ecology and the Environment* 6, no. 8 (2008): 439–47.

Flanagan, Owen. *Consciousness Reconsidered*. Cambridge, MA: MIT Press, 1992.

Foley, Jonathan A., Ruth DeFries, Gregory P. Asner, et al. "Global Consequences of Land Use." *Science* 309 (July 22, 2005): 570–74.

Gazzaniga, Michael S. *Human: The Science Behind What Makes Us Unique*. New York: Ecco, Harper Collins, 2008.

Ghosh, Amitav. *The Great Derangement: Climate Change and the Unthinkable*. Chicago: University of Chicago Press, 2016.

Holderness, Mike. "Think of a Number." *New Scientist* 170 (June 16, 2001): 45.

Kareiva, P., R. Lalasz, and M. Marvier. "Conservation in the Anthropocene. Beyond Solitude and Fragility." *Breakthrough Journal* 2 (Fall 2011): 29–37.

Kareiva, P. and M. Marvier. "What Is Conservation Science?" *BioScience* 62 (2012): 962–69.

Keith, D.W. "Geoengineering the Climate: History and Prospect." *Annual Review of Energy and the Environment* 25 (2000): 245–84.

Kelley, Kevin W., ed. *The Home Planet*. Reading, MA: Addison-Wesley, 1988.

Launder, B. and J. Thompson. *Geoengineering Climate Change: Environmental Necessity or Pandora's Box?* Cambridge: Cambridge University Press, 2010.

Lynas, Mark. *The God Species: Saving the Planet in the Age of Humans*. Washington, DC: National Geographic, 2011.

McCloskey, J. M. and H. Spalding. "A Reconnaissance Level Inventory of the Amount of Wilderness Remaining in the World." *Ambio* 18 (1989): 221–27.

McKibben, Bill. *The End of Nature*. New York: Random House, 1989.

Meyer, Stephen M. *The End of the Wild*. Cambridge, MA: The MIT Press, 2006.

Millennium Ecosystem Assessment. *Living Beyond Our Means: Natural Assets and Human Well-Being: Statement from the Board*. Washington, DC: World Resources Institute, 2005.

Parson, E. A. and David W. Keith. "End the Deadlock on Governance of Geoengineering Research." *Science* 339 (2013): 1278–79.

Pennisi, Elizabeth. "Are Our Primate Cousins 'Conscious'?" *Science* 284 (1999): 2073–76.

Pennisi, Elizabeth. "Brain Evolution on the Far Side." *Science* 314 (October 13, 2006): 244–45.

Penrose, Roger. *The Road to Reality: A Complete Guide to the Laws of the Universe*. New York: Alfred A. Knopf, 2005.

Peterken, George F. *Natural Woodland: Ecology and Conservation in Northern Temperate Regions*. Cambridge: Cambridge University Press, 1996.

Potts, Richard. "Sociality and the Concept of Culture in Human Origins." In *The Origins and Nature of Sociality*, edited by Robert W. Sussman and Audrey R. Chapman, 249–69. New York: Aldine de Gruyter, 2004.

Rees, Martin. *Just Six Numbers: The Deep Forces That Shape the Universe*. New York: Basic Books, 2000.

Royal Society of London. *Reaping the Benefits: Science and the Sustainable Intensification of Global Agriculture*. London: Royal Society, 2009. http://royalsociety.org/Reapingthebenefits/.

Scott, Alwyn. *Stairway to the Mind: The Controversial New Science of Consciousness*. New York: Copernicus; Springer-Verlag, 1995.

Seielstad, George A. *Dawn of the Anthropocene: Humanity's Defining Moment*. Alexandria, VA: American Geosciences Institute, 2012. (A digital book)

Shellenberger, M. and T. Nordhaus. "Evolve: A Case for Modernization as the Road to Salvation." *Orion* 30, no. 1 (September/October 2011): 60–65.

Sophocles. *Antigone*. 1900. R. C. Jebb, *Sophocles: Plays. Antigone* (facsimile reprint: London: Bristol Classical Press, 2004), line 325, p. 60.

Soulé, Michael E. "Conservation Biology in the Twenty-first Century: Summary and Outlook." In *Conservation for the Twenty-First Century*, edited by David Western and Mary Pearl. Oxford: Oxford University Press, 1989.

Steffen, Will, Paul J. Crutzen, and John R. Mitchell. "The Anthropocene: Are Humans Now Overwhelming the Great Forces of Nature?" *Ambio* 26, no. 2 (2007): 614–21.

Tattersall, Ian. *Becoming Human: Evolution and Human Uniqueness*. New York: Harcourt Brace, 1998.

Tegmark, Max. "Is the Theory of Everything Merely the Ultimate Ensemble Theory?" *Annals of Physics* 270 (1998): 1–51.

Thomas, Lewis. *The Lives of a Cell*. New York: The Viking Press, 1975.

Thompson, A. "Radical Hope for Living Well in a Warmer World." *Journal of Agricultural and Environmental Ethics* 23 (2010): 43–59.

Thompson, A. and J. Bendik-Keymer, eds. *Ethical Adaptation to Climate Change: Human Virtues of the Future*. Cambridge, MA: The MIT Press, 2012.

Tomasello, Michael, Ann Cale Kruger, and Hilary Horn Ratner. "Cultural Learning." *Behavioral and Brain Sciences* 16, no. 3 (1993): 495–552.

United Nations Conference on Environment and Development (UNCED). *The Rio Declaration*. 1992. www.unesco.org/education/pdf/RIO_E.PDF.

Venter, J. Craig, et al. "The Sequence of the Human Genome." *Science* 291 (February 16, 2001): 1304–51.

Victor, D. G., M. G. Morgan, J. Apt, J. Steinbumer, and K. Ricke. "The Geoengineering Option: A Last Resort Against Global Warming?" *Foreign Affairs* 88, no. 2 (March/April 2009): 64–76.

Wapner Paul. *Living Through the End of Nature: The Future of American Environmentalism*. Cambridge, MA: MIT Press, 2010.

Ward, Peter D. and Donald Brownlee. *Rare Earth: Why Complex Life Is Uncommon in the Universe*. New York: Copernicus; Springer-Verlag, 2000.

Waters, Colin N., et al. "The Anthropocene Is Functionally and Stratigraphically Distinct from the Holocene." *Science* 351 (2016): 137. http://dx.doi.org/10.1126/science.aad2622.

Wilkinson Bruce H. and Brandon J. McElroy. "The Impact of Humans on Continental Erosion and Sedimentation." *Geological Society of America Bulletin* 119 (2007): 140–56.

Wilson, Edward O. *On Human Nature*. Cambridge, MA: Harvard University Press, 1978.

Yoxen, E. *The Gene Business: Who Should Control Biotechnology?* New York: Harper and Row, 1983.

Index

Note: Page numbers in *italic* indicate a figure, and page numbers followed by an 'n' indicate a note on the corresponding page.

Printed in the United States
by Baker & Taylor Publisher Services